THE CONTROVERSY OF CLIMATE CHANGE

Why Are Politicians Trying to Scare Us?
The Truth Unveiled

Joseph Jablecki, PhD

ISBN: 978-1-6847-0929-8 (sc)
ISBN: 978-1-6847-0930-4 (hc)
ISBN: 978-1-6847-0928-1 (e)

Library of Congress Control Number: 2019913024

Lulu Publishing Services rev. date: 11/26/2019

This book is dedicated to

Sarah

Theresa

Jason

"In the beginning God created the heavens and the earth" (Genesis 1:1 King James Version). God is still doing so. It is a process that has gone on for eons and will continue to do so. The planets, sun, and other stars continue to change. As such, events beyond the control of man happen and will always continue to happen. Volcanos erupt, the earth quakes, and the seas rise and fall. The earth gets hot, temperate, cold, temperate, hot, temperate, cold, temperate …

CONTENTS

PREFACE

Global warming, climate change—whatever one chooses to call it—is real. Are human beings the cause? Perhaps, but certainly not to any significant extent. Can human beings stop it from happening? Absolutely not!

As you read this short book, you may notice occasional references to political and ideological commonality. It would be impossible to adequately cover the topic without some reference to such. Because, when it comes to belief, Republicans/conservatives differ greatly from Democrats/leftists.

While I was attending college for my PhD, my professors stressed that science is "the search for truth." Searching the internet for the definition of science, one finds many, but the consensus would be "the intellectual and practical activity encompassing the systematic study of the structure and behavior of the physical and natural world through observation and experiment." In this work, I prefer we be on a quest for truth.

At this point, you know I possess a doctorate, so let me expand a little on my qualifying background. My initial baccalaureate degree is in marine biology. A decade later I obtained a master's degree in environmental health science from Tulane University. Several years later I obtained a master's degree in business administration and then my terminal doctorate in environmental epidemiology. Over the past few years, I have become well read on the current theories and hypotheses concerning climate change. My conclusion reached is that it is happening, but people are not causing it. Further, people cannot stop it from happening.

Earth's climate is cyclical. It has gone through cycles for billions of years and will continue to do so, regardless of what humanity wants.

The purpose of this short book is to share with the population at large the scientific facts about the world's warming and cooling processes. Hopefully, this will end needless worry. Most people who profess that we need to change how we live to "save the planet" actually say these things to greatly increase their wealth. They do not really believe what they say.

What is presented in this book will in no way dispute that climate change is real. I concede now—climate change is real!

What I will dispute is that human activity is the cause of climate change. Once you have read this book, you will agree. So, if human activity is not the cause of climate change and climate change is happening, then what is causing it? After you have read this book, you will know and understand the cause.

To those who helped me with the editing—Michael F., Sarah H., and Betsy B.—thank you so much!

To those who gave me inspiration for writing the book—Al Gore and Alexandria Ocasio-Cortez—thank you.

To those who gave me permission to use data in this work, I also express my thanks.

CHAPTER ONE

Climate Change: The Myth

Let us first explore the myth being propagated about global climate change—that human beings are causing it, and therefore, human beings can stop it from happening. This is a myth. Why a myth? No truly scientific evidence exists that humans contribute to the course of climate change to any significant degree. Further, even if it were true that human beings are causing it, humans cannot stop it.

You may not be familiar with the myth proclamation.

It starts with an irrefutable assumption proclaiming scientific consensus, which implies that one cannot argue the point. In reality, no truly scientific evidence exists that humans contribute to the course of climate change to any significant degree. The myth contends the earth is getting warmer, mostly due to human activity. So, one might conclude that all we have to do is stop what humans are doing. But let us give them the premise that human beings are the cause of Climate Change. Can we be so arrogant as to profess we are so powerful we can actually control Climate? No, of course not as there are far more significant factors at play.

Why do people support this myth? Most people want to control their destinies. They want to believe it's humanly possible to control the planet's climate and environment. Simultaneously, other people are making a great deal of money by perpetuating the myth that humans cause global warming.

Al Gore, former vice president, is said to have a net worth in excess of $300 million dollars. Other than taking credit for creating the internet, he has never built or created anything. So how did he become so rich? Well, first, he is a Democratic politician, and many Democratic politicians become rich. For that matter, most politicians become rich. But in the case of Mr. Gore, he accomplished it by instilling fear into the population.

He is now chairman of Generation Investment Management, chairman of the Climate Reality Project, senior board member of Apple, Incorporated, chairman of the Alliance for Climate Protection, and a senior partner at Kleiner Perkins—all because of his political clout. Wait, there's more. In 2007, Al Gore was awarded the Nobel Peace Prize along with the members of the Intergovernmental Panel on Climate Change. He has also received an Academy Award for his trepid movie *An Inconvenient Truth*, a Grammy award for the narration of his book *An Inconvenient Truth*, an Emmy award, a Webby award for his website, and, yes, here it is: he was runner-up for Person of the Year of the ultraleft, fear-mongering *Time* magazine.

So what is his *An Inconvenient Truth* all about? Gore published a book and made a movie on the subject in 2006. He said things like "humanity is sitting on a time bomb"; "global warming is really not a political issue, so much as a moral one"; and "we have just ten years to avert a major catastrophe that could send our entire planet's climate system into a tailspin of epic destruction." Yet as of today, the "time bomb" has yet to explode, global warming remains a political issue, and the "ten years" have long passed without the "major catastrophe."

Gore discusses the "scientific opinion" on global warming, yet it is not scientific. Opinion, yes. But the opinion is fanciful garble. He says he believes global warming will destroy the planet if human-generated greenhouse gases are not significantly reduced. Does he really believe that? I think not. But the fear he has instilled has generated a fortune for himself.

So how does Gore sell this snake oil? He starts by explaining, with fifth-grade vocabulary, that there is a measurable correlation between increases in carbon emissions and increases in global temperature. He gives the reader a simple explanation of the basic, foundational "theories" on how climate change works. These theories are his. He then goes into the horrific devastation these temperature changes will produce in terms of weather changes, such as more frequent and massive storms as well as numerous worldwide natural catastrophes. In the movie version, Gore displays photographs in a slide presentation that are graphic and scary.

Gore's explanation of the effects of greenhouse gases on the temperature of Earth encompasses twenty-five pages of the book. Then, for no particular reason, he talks about his old Harvard professor Roger Revelle. Book filler is all one should make of it. He goes on to describe the harm receding glaciers cause in places like Mount Kilimanjaro and Glacier National Park. Then he tells us about his son being struck by an automobile and how he got into politics—more book filler having nothing to do with climate change. In fact, if you read his book, you will find a constant mingling of Gore's personal and professional lives splintered with occasional alarming statements about climate change.

When he does talk about climate change, Gore always

stresses "scientific evidence." At one point, he states there will be increased flooding around the world and then, almost in the next sentence, horrible droughts. Which is it? Well, it is both. But that is not unusual as in the United States in the past year—floods in the Midwest and droughts in the West.

He uses up a few pages with the story of his sister's battle with lung cancer, pointing out that people smoke because they don't understand the link between smoking and deadly cancer—because the cancer can grow slowly over time. Really? People don't understand that? To me, the most interesting part of the book is Gore's turmoil between his life on the farm and city life.

But Gore ends his doom and gloom with a rally cry to make us all feel better about the future. He calls us into action! He gives us specific things we can do to slow the rate of carbon increases and, accordingly, slow climate change. He asks us to take a more active role in politics and to make smarter choices in how we spend our money, and he shows ways to mitigate purchasing items that will negatively affect the environment. Who can argue with that logic?

Liberals eat this stuff up because they always want to use government action to achieve ill-conceived goals. But are these people liberal? No. Liberal is not a good descriptive term for them; the word liberal is derived from the Latin liber, meaning "free," which is also the root of liberty, the state of being free. The Oxford Dictionary defines liberal as "willing to respect or accept behavior or opinions different from one's own; open to new ideas." Today's liberals are anything but liberal. Rather, they are closed minded and incapable of respect and force their ideas on others. So from this point forward, I will refer to them as leftists.

Most likely without knowing it, Gore is right about a lot of things. He just has the wrong understanding of the how and why.

- The overall average temperature of the earth is rising, just not as fast as Gore states. Actually, over the past two hundred years, the average global temperature has risen about two degrees Fahrenheit.
- Most likely, before Earth begins to cool again (and it will), the majority of life as we know it will perish.
- There has been and will continue to be a measurable increase in the frequency of major storms and their magnitude.
- After the warming phase, Earth will cool again to what we currently refer to as "normal." When that happens, some of the species-specific life we have today will return. But there will also come new life forms.
- And Gore is right about the ice melting. It will melt. All of it.

Most likely not because he needed the money—I think—Gore produced an updated version and titled it *An Inconvenient Truth Sequel*. It seems the intent of this update is to scare you even more.

Al Gore, of course, is not the only sermonizer for human-caused climate change (HC³). Take, for example, the unconscious and factually ignorant, newly elected US representative for New York's fourteenth congressional district, Alexandria Ocasio-Cortez, or AOC as she is affectionately known. AOC has some vocal opinions. She claims credit for defining what she calls a Green New Deal—a massive policy package that would remake the US economy and eliminate all US carbon emissions. A lofty and impossible dream. She has also said that climate change is "our World War II," and "the world will end in twelve years." She should have a talk with Ted Danson to get an understanding of the folly in making such outlandish public statements. (If you are unaware of Mr. Danson's proclamation, in 1988, he predicted, and I paraphrase, "The oceans are gone with no hope of saving. In just ten years all will be lost, and you'll be drinking your neighbor's urine.")

But Ted is not alone; many others, including ecologist Kenneth Watt; biologist George Wald; biologist Paul Ehrlich; chief organizer of Earth Day, Denis Hayes; North Texas State University professor Peter Gunter; and Senator Gaylord Nelson have failed to predict Earth's demise.

While there certainly is an ever-growing consensus that our climate is changing, fortunately, most people do not believe the world will end in just twelve years. In fact, beliefs about the causal factors are mixed among the public. A recent observation cited by Joel Hartter,[1] with extensive references on the work conducted by Hamilton,[2] indicates causal beliefs are solidly influenced by cultural, political, and identity-driven views (see figure 1). Point of note: 40 percent of eastern Oregon residents believe the scientific consensus that human activities are now changing the climate, while 44 percent of eastern Oregon residents believe climate change is happening now but is naturally occurring without human activity. In the 2011 survey, more people believed climate change was happening, but nature was the cause. However, in just three years (2014), an unexpected paradigm shift occurred: the majority of respondents believed climate change was the result of human activities. Perhaps this is because the people of Oregon are primarily politically leftists.

---

[1] Joel Hartter et al., "Does It Matter If People Think Climate Change Is Human Caused?" *Climate Services* 10 (April 2018), https://doi.org/10.1016/j.cliser.2017.06.014, 53–62.

[2] L. C. Hamilton et al., "Forest Views: Shifting Attitudes toward the Environment in Northeast Oregon," (Carsey Institute, 2012); L. C. Hamilton et al., "Rural Environmental Concern: Effects of Position, Partisanship, and Place," *Rural Sociology* (2014); L. C. Hamilton et al., "Tracking Public Beliefs about Anthropogenic Climate Change," *Public Library of Science* 10, no. 9 (2015); L. C. Hamilton et al., "Wildfire, Climate, and Perceptions in Northeast Oregon," *Regional Environmental Change* (2016).

Belief AboutClimate Change

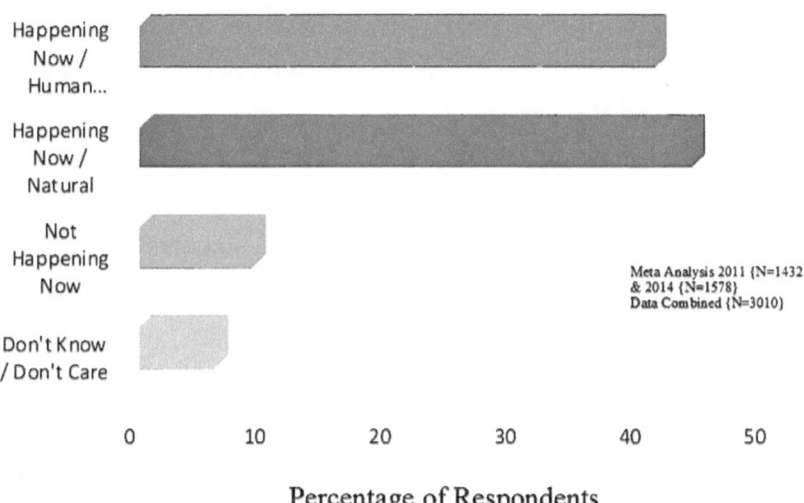

Meta Analysis 2011 {N=1432}
& 2014 {N=1578}
Data Combined {N=3010}

Percentage of Respondents

Figure 1: Graphic by Jablecki via meta-composition of depictions in: Joel Hartter, Lawrence C. Hamilton, Angela E. Boag, Forrest R. Stevens, Mark J. Ducey, Nils D. Christoffersen, Paul T. Oester, Michael W. Palace, "Does it matter if people think climate change is human caused?", Science Direct, Climate Services, Volume 10, April 2018.; Hamilton, L.C., , Hartter J., Stevens,, F., et al. "Forest views: shifting attitudes toward the environment in northeast Oregon", Carsey Institute, 2012.; Hamilton, L.C., Hartter J., Safford, T. G., Stevens, F. R., "Rural environmental concern: Effects of position, partisanship and place", Rural Sociology, 2014.; Hamilton, L.C., Hartter J., Lemcke-Stampone, M., et al., "Tracking public beliefs about anthropogenic climate change", Public Library of Science Vol 10 (9), 2015.; Hamilton, L.C., Hartter J., Keim B.D., et al., "Wildfire, climate, and perceptions in northeast Oregon", Reg. Environ. Change, 2016.

The last two response groupings indicated climate change "is not happening now" and "I just don't know or care." These are people not involved in civic responsibility. Even so, the bottom line of the study indicates that acceptance of anthropogenic climate change is divided along political party lines.

Considering this study was conducted in a West Coast state, one could assume all "blue" states would follow with a similar outcome. One could also assume that "red" states would predominantly respond "yes, climate change is happening, but humans are not the cause."

Another recent study[3] administered a twenty-question survey to Texas Tech University students. The questionnaire was titled "Online Survey to Collect Opinions about Factors Influencing Global Warming." Findings indicate most people are woefully uninformed as to the likely cause, and impossible mitigation, of climate change. Again, the analysis indicated that the opinions and beliefs of these students were drawn along political lines. Democrat/leftist-leaning students had a stronger belief association than either the Independent group or the Republican group (see figure 2).

An important consideration for those scientists who may be reading this: Hartter et al. (2018) ran statistical analysis on their data using logistic regression (an analysis conducted on binary [two elements] data to determine if the answer found is true or not).

---

[3] Jonathan E. Thompson, "Survey Data Reflecting Popular Opinions of the Causes and Mitigation of Climate Change," *Data in Brief* 14, (October 2017): Science Direct; L. C. Hamilton et al., "Tracking Public Beliefs about Anthropogenic Climate Change," *Public Library of Science* 10, no. 9 (2015); L. C. Hamilton et al., "Wildfire, Climate, and Perceptions in Northeast Oregon," *Regional Environmental Change* (2016).

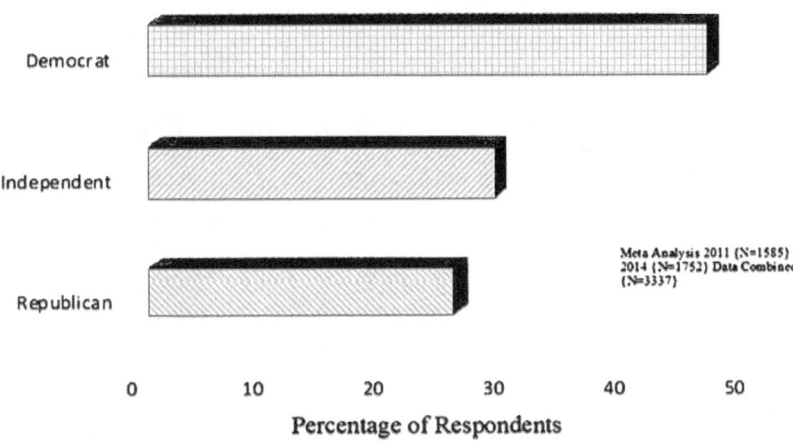

Climate Change Caused by Humans

Meta Analysis 2011 (N=1585) &
2014 (N=1752) Data Combined
(N=3337)

Percentage of Respondents

Figure 2: Graphic by Jablecki via meta-composition of depictions in: Thompson, Jonathan E, "Survey data reflecting popular opinions of the causes and mitigation of climate change", Science Direct, Data in Brief, Volume 14, October 2017.; Hamilton, L.C., Hartter J., Lemcke-Stampone, M., et al., "Tracking public beliefs about anthropogenic climate change", Public Library of Science Vol 10 (9), 2015.; Hamilton, L.C., Hartter J., Keim B.D., et al., "Wildfire, climate, and perceptions in northeast Oregon", Reg. Environ. Change, 2016.

They found that, for all questions answered in the survey, the result was statistically significant (p < 0.001). Such a resultant *p* value indicates the result is truth. The *p* stands for predictive. The *p*-value is a number between 0 and 1, which is used to test the result of an analysis. In the test, two theories (binary) are possible. One being the null (not contributing) hypothesis, the other being the alternative (it is true) hypothesis. A small resultant *p*-value (<0.05) indicates strong evidence against the null hypothesis, so you reject the null hypothesis and accept the alternative. A large *p*-value (> 0.05) indicates weak evidence against the null hypothesis, so you fail to reject the null hypothesis. This would mean your result is more likely not truth.

Another irrational factoid includes the myth, "Forest fires are increasing as a result of climate change."[4] In truth, no, they are not. If anything, they may actually be decreasing in frequency. Why the myth is so accepted is that although they are not increasing in frequency, they are increasing in size. This is particularly true in West Coast blue states. Why? No deforestation, so there is too much fuel.

"Electric powered transportation will save the planet." That is what most leftist leaning people tell me. In San Antonio, where I live, the city is about to launch a new Climate Action and Adaption Plan.[5] This great idea will "stop adding more greenhouse gases to the atmosphere by 2050." The proponents of the Climate Action and Adaption Plan want to use taxpayer dollars to install lots of electrical charging stations throughout the city, and they want to subsidize low-income people so they can buy electric cars.

Here's the problem with "green" thinking. Electricity is not

[4] Pierre-Louis, K., Popovich, N., "Climate Change Is Fueling Wildfires Nationwide, New Report Warns", *New York Times,* Nov. 27, 2018.
[5] SA Climate Ready, City of San Antonio Climate Action and Adaptation Plan (CAAP), Feb. 16, 2019.

magic. Electricity has to be produced. Some is created by the power of gravity, as in hydroelectric power produced as water falls through turbines turning rotors and generating power. Some is created by solar, some by wind, some by nuclear fission, and a great deal by burning natural gas or coal. San Antonio's power company City Public Service (CPS) uses a variety of sources for its power development, including purchasing from home owners their excess solar-generated power. But 45 percent of the electricity produced and sold by CPS comes from the burning of natural gas and another 8 percent from burning coal. So what is saved from the "carbon footprint"? Nothing.

San Antonio, Texas, is a typical large US city, meaning it is run by leftists (progressives). So the city managers developed an eighty-four-page plan called "Climate Action and Adaption Plan." To justify the high cost of this action, they cited "undisputable scientific studies"; the city's alignment with, and support of, the Paris Climate Agreement; and further stated, "San Antonians are already feeling the impacts of climate change with wildfires, powerful storms, and intense heat." According to their studies, San Antonio will have an average temperature four degrees higher in just twenty-one years. Science says this is not true, as you will read later in the book.

City leaders proudly proclaim, "San Antonio is climate ready." Sounds impressive, but what the hell does that mean? The development of the eighty-four-page plan cost a half million dollars put up by City Public Service, the power company. So, all CPS has to do is raise the cost per kilowatt hour for a while, and they recoup the money from the public. The plan is a lot of feel-good stuff that is totally impracticable at nearly $6,000 per page. I'd like to say none of what is in the document is true, but perhaps something in there may be.

San Antonio is but a single drop in an ocean of water. Every leftist-run city in the United States has such a plan. They tout the Paris Agreement as if it were the New Testament, but there is absolutely nothing concrete in that document. Read on and you will discover the truth about climate change. It is real, but there is nothing any human can do to stop it, except perhaps pray, but I doubt that either. Climate change has been going on since the dawn of time, and I see no reason for God to end it now.

So, here's a short recap of the myths and why they perpetuate: climate change is a new phenomenon that has never happened before. It is happening now because the population of Earth is too large, and Earth cannot sustain it as such. Humans are polluting the atmosphere with vast amounts of greenhouse gases. If we humans do not correct this, the climate will change in just a few years, to the point where animal life will no longer be able to survive on the planet. Humans have the power to stop climate change.

The proponents purport (and may even believe) humans are the sole master of our environment. We cause things to happen, and we can fix anything. Now is the time to absorb what is really happening. By the end of this book, you will understand the truth! Read on.

CHAPTER TWO

The Truth about Our Atmosphere

Factors that make Earth an inhabitable planet are numerous, but most important are the terrain, proximity, and atmosphere. Earth's terrain is primarily rocky, which absorbs little energy and reflects radiation. Earth's proximity is the perfect distance from the host star (sun). Any closer—too much solar radiation, and any farther—not enough solar radiation. While several planets and some moons in our solar system have atmospheres, Earth's is just right for supporting life.

According to the proponents of human-caused climate change, human activity-caused pollution of Earth's atmosphere is the reason for climate change. These human-caused climate change proponents convened the UN Framework Convention on Climate Change (UNFCCC), which led to the Paris Agreement of April 2016.[6] This agreement was to "reduce the threat of climate change by keeping a global temperature rise during this century to below two degrees Celsius above preindustrial levels." Okay,

---

[6] UNFCCC, Paris Climate Change Agreement, Apr. 22, 2016.

but how? Their stated method was "to appropriate financial flows, develop a new technology framework, and an enhance the capacity building framework."[6]

I would like you to read that again: "to appropriate financial flows, develop a new technology framework, and an enhance the capacity building framework." Perhaps read it several more times. What a bunch of meaningless mumbo jumbo! The only thing that means anything is "appropriate financial flows," which means to get as much money as possible. To top it off, the United States was to handle the lion's share burden of the financial contribution. Yet as of this writing, China continues to burn 49 percent of the world's coal consumption worldwide. The United States burns about 11 percent of the consumed coal per year.[7]

Most people believe the United States is the world's biggest polluter. It's not. While the United States does contribute significantly to the world's CO_2 level, it's not the biggest contributor of CO_2 either (see figure 3). In fact, today, the United States is one of the global nations that has the lowest levels of air pollution.[8] Yet the United States does the most to reduce air pollution globally.

In terms of metric tons of CO_2 emission per capita, in 1960 the United States ranked third in the world among the 241 nations/ regions providing data (see appendix A). In the resurvey of 2014, the US emissions level was nearly exactly what it was in 1960. However, the United States had dropped from third to eleventh. Very important to note—data was self-reported.

[7] Today in Energy, U.S. Energy Information Administration, May 14, 2014. Everything You Think You Know About Coal in China Is Wrong. Energy and Environment. Center for American Progress. May 15, 2017.

[8] WHO, Ambient air pollution: A global assessment of exposure and burden of disease, 2016, ISBN: 9789241511353, Key World Energy Statistics, The International Energy Agency data, analysis, and solutions on all fuels and technologies, https://www.iea.org/statistics/kwes.

Countries with the Highest AIR POLLUTION

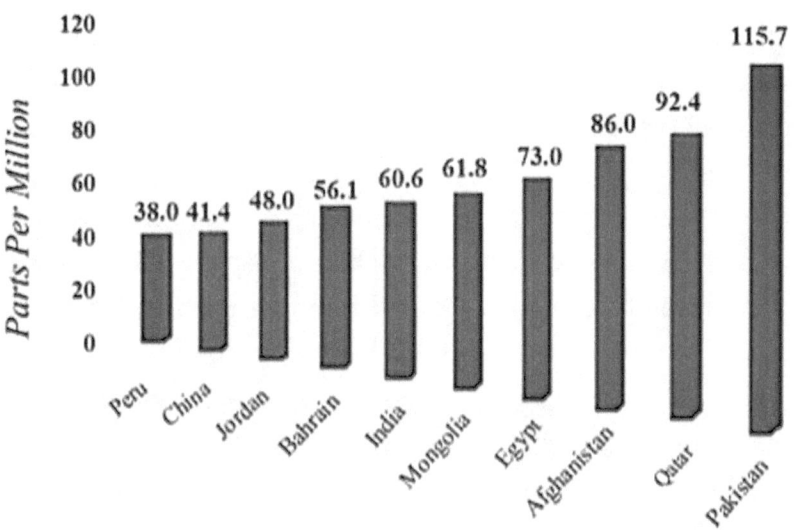

Countries with the Lowest AIR POLLUTION

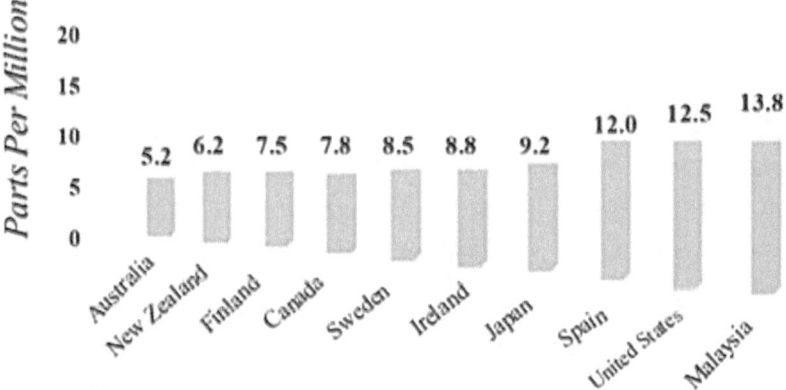

Figure 3: Adapted by Jablecki using WHO data from "Ambient air pollution: A global assessment of exposure and burden of disease, 2016, ISBN: 9789241511353" and IEA data from "Key World Energy Statistics, The International Energy Agency data, analysis, and solutions on all fuels and technologies, https://www.iea.org/statistics/kwes/"

Life on Earth depends on energy coming from the sun. About half the light reaching Earth's atmosphere passes through the air and clouds to the surface, where it is absorbed and then radiated upward in the form of infrared heat. About 90 percent of this heat is then absorbed by the greenhouse gases and radiated back toward the surface, which has been warmed to a life-supporting average of fifty-nine degrees Fahrenheit. That is good. We need this warmth.

So is there some sort of increase in the amount of radiated heat coming back to Earth from the atmosphere? This is what the HC[3] pundits tell us. How do we know that changes in the sun aren't to blame for current global warming trends?

Since 1978, a series of satellite instruments have directly measured the energy output of the sun. The satellite data show a slight drop in solar irradiance. (Solar irradiance simply means "shining brightly" but is also a measure of the amount of energy the sun gives off.) As a measure, solar irradiance is the flux of radiant energy per unit area over this time period. The data tell us that since 1978, there has been only a slight drop in energy, not a slight rise, so then the sun doesn't appear to be responsible for the warming trend observed over the past several decades.

Longer-term estimates of solar irradiance have been made using sunspot records and other so-called "proxy indicators," such as the amount of carbon in tree rings. The most recent analyses of these proxies indicate that solar irradiance changes cannot plausibly account for the estimated 1 percent rise during the twentieth century's warming.

So if the sun is not producing more heat, then it must be the atmosphere. Well, many have come to this conclusion, but it is a false rationale. Earth's atmosphere isn't that much different from what it's been for the past thousands of years, going back as far as we can with geological data.

Billions of years ago, volcanoes spewed gases that combined with the water that had accumulated from space via gravity, and Earth's atmosphere was formed. Volcanoes still spew gases today, and water still enters the earth from space. The difference today is volcanic eruptions are far fewer than in the past, primarily because back in the day, Earth's crust was much thinner. Also, mantle convection has been slowing down for a long time.

Our atmosphere is perfect for life. If it were any thicker, the planet's surface would be too hot. If it were any thinner, the planet's surface would be too cold. However, the argument could be made that our atmosphere is getting thicker from greenhouse gas production by humans, and hence the surface is getting hotter. A good argument, but there is no supporting evidence.

Let us now understand our atmosphere better. It is around three hundred miles thick, yet the preponderance of gases lies within the first ten miles. In full measure, it is an overall mixture of nitrogen (78 percent), oxygen (21 percent), and other gases (1 percent). Included in the other gases are the "evil" greenhouse gases. But they are not evil at all and, alternatively, are beneficial.

The atmosphere is divided into seven somewhat arbitrary layers or zones (see figure 4):

- troposphere
- stratosphere
- ozone layer
- mesosphere
- thermosphere
- ionosphere
- exosphere

However, in most literature, only five layers are recognized: troposphere, stratosphere, mesosphere, thermosphere, and exosphere. Our atmosphere thins out significantly with each higher layer until the gases finally leave the atmosphere, dissipating into space.

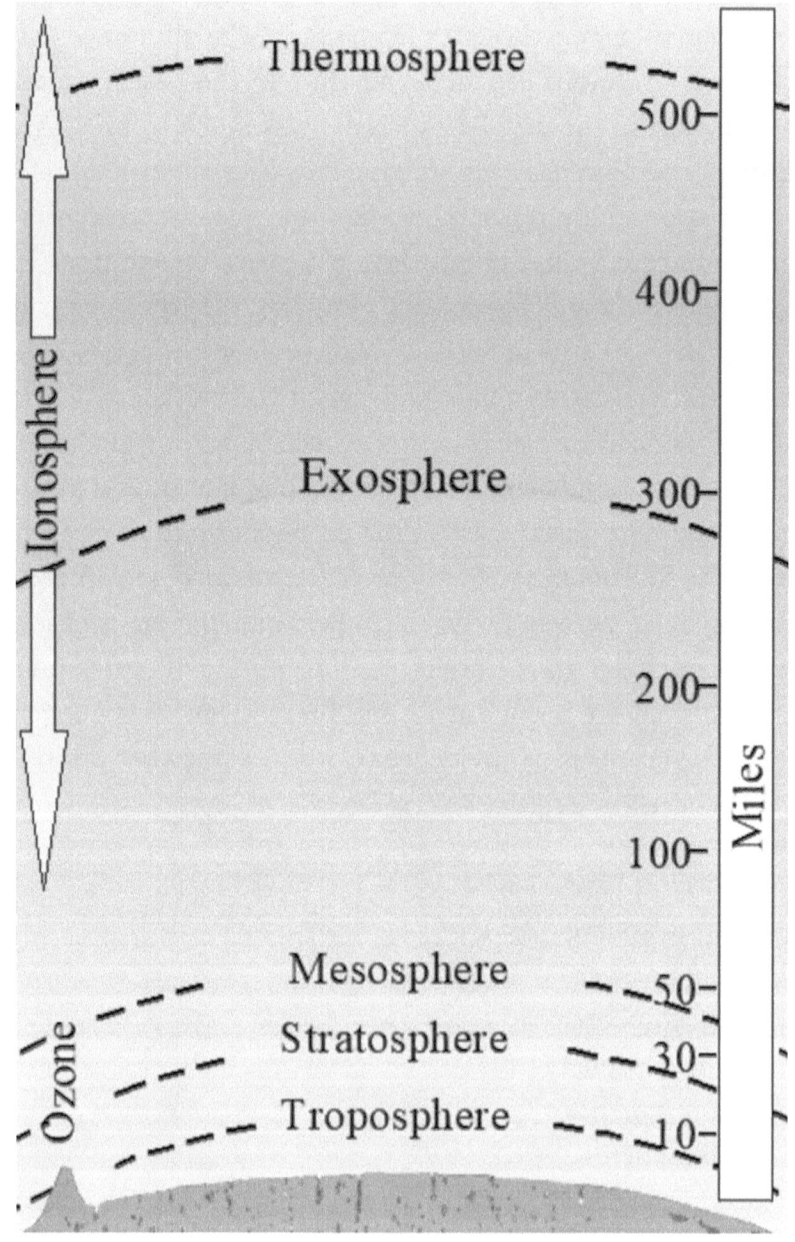

Figure 4: Source: Graphic by Jablecki | Earth's seven atmosphere Layers or Zones.

The troposphere lies from the earth's surface to around ten miles up. It contains significantly more gases than any other layer. The air temperature is warmer at the ground and drops at a rate of approximately 3.5 degrees Fahrenheit per thousand feet as altitude increases to 36,000 feet. At that altitude, the temperature is relatively constant at approximately minus sixty-five degrees Fahrenheit and remains such until reaching the stratosphere, where the temperature radically drops. The reason for the warmer temperature at the lower troposphere is because the more gas, the more heat retention. Also, the water vapor and dust in the atmosphere are contained in this layer, which is why clouds form here.

The stratosphere is the layer just above the troposphere and extends up to about thirty-one miles above ground. Ozone is abundant in the stratosphere, which helps heat the atmosphere while also absorbing harmful radiation from the sun. The air in this layer is incredibly thin and dry.

The mesosphere is the next layer up, starting at the top of the stratosphere (thirty-one miles) and extends to about fifty-three miles above the earth's surface. The highest portion of the mesosphere is termed the mesopause, where the temperature averages minus 130 degrees Fahrenheit. The gases in this layer are incredibly thin and contribute little to the warming of the earth's surface.

The thermosphere extends from the top of the mesopause (about fifty-six miles) up to the highest defined layer of the atmosphere, around three hundred miles. Here is an oddity you may have trouble comprehending: because there is hardly any gas in this layer, the sun's radiation is unbridled. Here the temperature can reach 2,700 degrees Fahrenheit. Although the thermosphere is considered part of Earth's atmosphere, air density is nearly nonexistent.

The ionosphere is poorly defined and is actually interposed within the thermosphere and mesosphere. It extends from the mesosphere to as much as five hundred miles above Earth's surface. This layer contains a very high concentration of ions and free electrons and is able to reflect energy waves such as radio waves.

The exosphere is the highest layer. Gases are so sparse and protracted that differentiation with absolute space is not demarcated. This is the layer that merges with space. It is composed of widely dispersed particles of exclusively hydrogen and helium.

What about the ozone layer? Well, it is kind of a layer within layers, which is why I saved it for last. If you're like most people, you have heard ozone (O^3) is horrible! Not so. The ozone layer is also known as the ozone shield, and a shield is a good thing. Most of the ozone shield is contained within the lower portion of stratosphere, and it is this shield that absorbs most of the solar ultraviolet radiation. The thickness of this layer varies both seasonally and geographically. For the most part, the ozone layer contains fewer than ten parts per million (ppm) of ozone, while the average ozone concentration in Earth's atmosphere as a whole is less than 0.3 ppm.

So what is the deal with these greenhouse gases? I hear they are the worst thing happening these days. I also hear we must limit their production and reduce each and every person's "carbon footprint." Do you hear those same things? Read on.

CHAPTER THREE

Greenhouse Gases

Now, before I explain the real cause of climate change, let us discuss the greenhouse effect. As I have previously said, a great many people consider the greenhouse effect to be the cause of climate change. And they also think people are the primary cause of the excess greenhouse gases. Some call it man-caused climate change, while others refer to it as human-caused climate change, or HC^3.

To paraphrase the general consensus: Human activities are changing Earth's natural atmosphere's greenhouse protection. The burning of fossil fuels like coal and oil is the cause producing an increase in the concentration of carbon dioxide. This occurs when burning coal or oil carbon combines with oxygen to form CO_2.

Somewhat, but this is not the cause of climate change.

The greenhouse effect (see figure 5) contends that the increase of certain gases, mostly carbon dioxide (CO_2), causes the earth to become warmer than it used be. Many consider the premise to be scientific fact. And to some degree, it is a fact. But it's a meaningless fact, because the contribution to greenhouse gas by human activity is insignificant. So then, are people responsible for the increase in these greenhouse gases? To some degree, yes. But not to the extent of changing the climate.

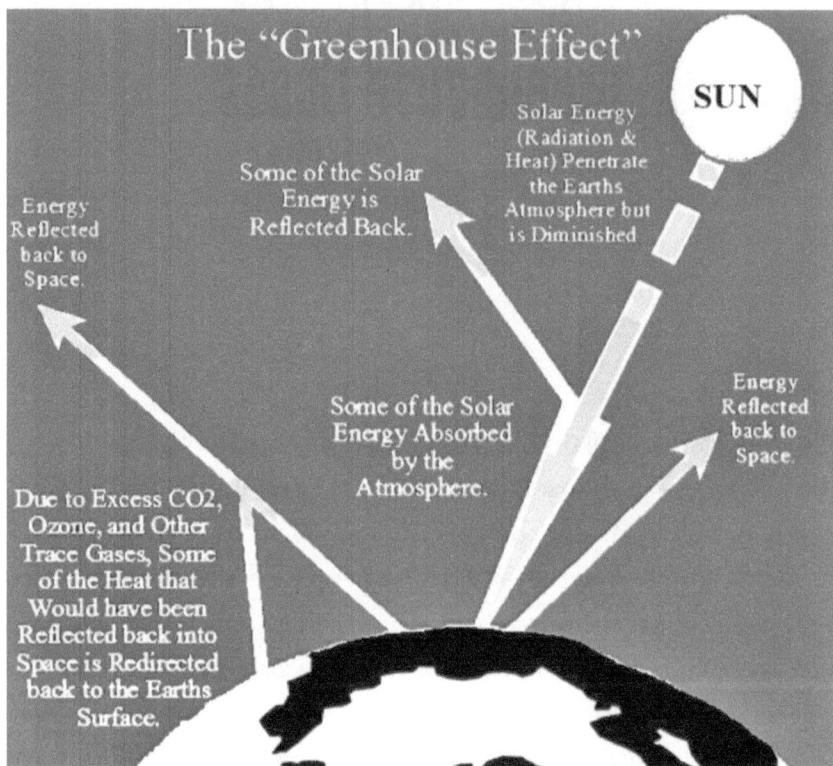

Figure 5: Source: Graphic by Jablecki via meta-composition of depiction in: "Earth's energy budget",with incoming and outgoing radiation. Satellite instruments (CERES) measure the reflected solar and emitted infrared radiation fluxes. The energy balance determines Earth's climate. Wikipedia | Energy flow between the sun, the atmosphere and earth's surface; Greenhouse effect, https://en.wikipedia.org/wiki/Earth%27s_energy_budget#/media/File:The-NASA-Earth's-Energy-Budget-Poster-Radiant-Energy-System-satellite-infrared-radiation-fluxes.jpg, File:The-NASA-Earth's-Energy-Budget-Poster-Radiant-Energy-System-satellite-infrared-radiation-fluxes.jpg, Public Domain, Created: 19 April 2014.

The primary greenhouse gases in Earth's atmosphere are water vapor (H_2O), carbon dioxide (CO_2), methane (CH_4), nitrous oxide (N_2O), and ozone (O_3). The two touted as being increased by humans are carbon dioxide and ozone. The average ozone concentration in Earth's atmosphere as a whole is about 0.000004 percent by volume. Now, that figure is often separated into stratospheric "good" and tropospheric "bad" ozone. The current estimate is stratospheric ozone (good) is 0.00006 percent and tropospheric ozone (bad) is 0.000001 percent. The average of the two is 0.0000035 percent. The figure is rounded up, which is how they come up with the figure 0.000004 percent by volume. Still, even rounded up, it's not much, is it?

That 0.000001 percent tropospheric ozone equates to ten parts per billion (ppb) or only 0.01 parts per million (ppm). Or, for every one million gas molecules, less than one is O_3. Ozone is a very important gas to the survival of life because O_3 absorbs UV light from the sun. And, yes, in doing so, temperature increases. But it is a valuable tradeoff.

Ozone's role in the enhancement of the greenhouse effect is indeterminable. It is impossible to know what the concentrations of O_3 were in the previous history of the world. Measurements of past ozone levels more than forty years in the past are not unavailable. When reviewing the past forty years of data we do have, the conclusion is again indeterminate. There has been no significant change in ozone in the atmosphere over the past forty years. I repeat—*there has been no significant change!*

The other primary greenhouse gas of concern is carbon dioxide. Concentrations of CO_2 in the atmosphere were once as high as five thousand ppm during the Cambrian and Ordovician periods (around five hundred million years ago) and as low as five ppb during the Quaternary glaciation of the last two million years (see figure 6).

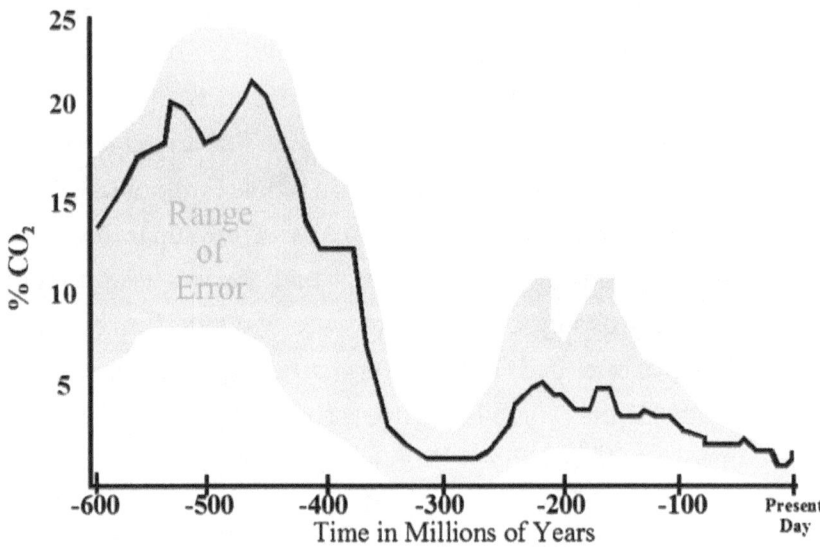

Figure 6: Source: Graphic by Jablecki via meta-composition of depictions in: "Science Direct", <u>CO2-forced climate thresholds during the Phanerozoic</u>, Dana L. Royer, 30 November, 2005; "Science", <u>The Rise of Plants and Their Effect on Weathering and Atmospheric CO2</u>, Robert A. Berner, Apr. 25, 1997; "Time Scavengers", <u>CO2: Past, Present, & Future</u>, Blog, January 14, 2019.

So, one may ask: "Up to what level of CO_2 is harmful to humans?" Current US Environmental Protection Agency (EPA) guidelines state that when the CO_2 level is one thousand to two thousand ppm, the air quality is low. A two thousand to five thousand ppm CO_2 concentration can cause problems (such as headaches, insomnia, and nausea) for people. From five thousand ppm and above, the presence of other gases in the air is altered, arising to a toxic atmosphere, very poor in oxygen, with fatal effects for humans as the concentration increases.

Yet carbon dioxide concentrations in today's atmosphere are minimal. The daily average concentration of atmospheric CO_2 is about four hundred ppm. Recently (April 2018), the average monthly level of CO_2 in Earth's atmosphere exceeded 410 ppm for the first time in the modern era. It will most likely increase, and the increase is probably due to human contribution. So will reducing the human component of CO_2 help stop global warming? No! Why? Because greenhouse gases are not the cause.

Let's make sure we're clear: these numbers are *global!* There are many "pockets" where CO_2 levels are much, much higher, and there are pockets where CO_2 levels are somewhat lower. These high pockets are industrial regions where substances such as coal, oil, and wood are burned. Also, areas around simmering volcanoes are high pockets.

Okay, let's reflect before we continue. Are we saying humans should pump CO_2 into the atmosphere indiscriminately? Of course not. It is our planet, and we should respect it. But will the minimal increase humans contribute to CO_2 affect global warming? No, absolutely not. It has no effect. Please read on.

So yes, the composition and content of greenhouse gases in the earth's atmosphere has changed over the past several billion years. Scientist claim that about five billion years ago, the earth's first atmosphere evolved from the vapor and gases that were expelled during degassing of the planet's interior. You do know that beneath the earth's crust is hot molten metal, nearly as hot as the surface of the sun? You know this, don't you?

Then, about four billion years ago, the earth's second atmosphere evolved from the condensation of water vapor, resulting in oceans of water in which sedimentation occurred. Elements emerged. About one billion years ago, early aquatic organisms such as blue-green algae began using the sun's energy to split the molecules of H_2O and CO_2 and recombine them into organic compounds and molecular oxygen (O_2). This solar energy conversion process is known as photosynthesis. Some of the photosynthetically created oxygen combined with organic carbon to recreate CO_2 molecules. The remaining oxygen accumulated in the atmosphere, touching off a massive ecological disaster with respect to early existing anaerobic organisms. As oxygen in the atmosphere increased, CO_2 decreased. This became the atmosphere Earth has today: composition of 78 percent nitrogen, 21 percent oxygen, and 1 percent other gases. Most of the other gas is argon, and then very trace amounts of carbon dioxide, neon, helium, methane, krypton, hydrogen, nitrous oxide, xenon, ozone, iodine, carbon monoxide, and ammonia. How much CO_2? About 0.004 percent!

In fact, carbon dioxide is an important trace gas in Earth's atmosphere. It's an integral part of the carbon cycle, a biogeochemical series of events in which carbon is exchanged between Earth's oceans, soil, rocks, and the biosphere. Plants and microbes utilize solar energy to produce carbohydrate from atmospheric carbon dioxide and water by the process of photosynthesis. Almost all other organisms depend on carbohydrate derived from photosynthesis as their primary source of energy and carbon compounds. Also, like ozone, CO_2 absorbs and emits infrared radiation and consequently is a greenhouse gas that plays a vital role in regulating Earth's surface temperature through the greenhouse effect.

So I have to agree with Al Gore and the other HC^3 group that over the past one hundred years, the level of CO_2 in the atmosphere has increased. But I must vehemently disagree that harm has resulted. Further, there is no scientific evidence this increase has been caused by human activities, particularly the burning of fossil fuels and deforestation. After all, we do know that scientific evidence establishes that CO_2 levels have risen and fallen when humans were not even alive.

Carbon dioxide concentrations have varied widely over Earth's 4.54-billion-year history. Restating to be clear: carbon dioxide is believed to have been present in Earth's first atmosphere, shortly after Earth's formation.

The second atmosphere, consisting largely of nitrogen and carbon dioxide, was produced by outgassing from volcanism, supplemented by gases produced during the late heavy bombardment of Earth by huge asteroids. A major part of carbon dioxide emissions was soon dissolved in water and built up carbonate sediments. The production of free oxygen by cyanobacterial photosynthesis eventually led to the oxygen catastrophe that ended Earth's second atmosphere and brought about Earth's third atmosphere (the modern atmosphere) 2.4 billion years before the present. Carbon dioxide concentrations dropped from four thousand parts per million during the Cambrian period about five hundred million years ago to as low as 180 parts per million during the Quaternary glaciation of the last two million years. Over the most recent geological period, four hundred thousand years ago, CO_2 levels were about the same as they are today (see figure 7).

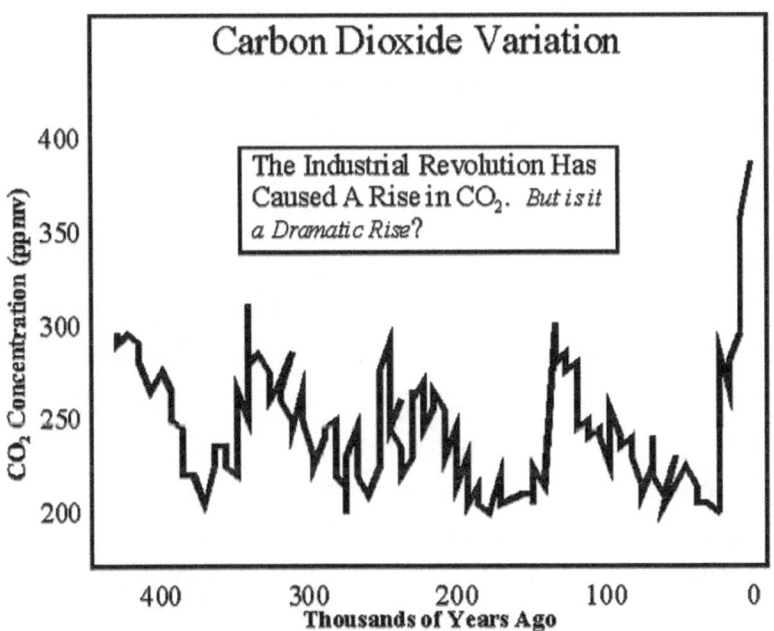

Figure 7: Source: Carbon dioxide 400kyr, Wikipedia | Variations in atmospheric carbon dioxide levels over the last 400,000 years.

However, during those four hundred thousand years, the CO_2 levels have continually waxed and waned, without human contribution.

Now, when Al Gore says the CO_2 level has doubled (even though it hasn't), that would be only 0.08 percent. Granted, he claims to have created the internet with his Al-Gore-Rhythm, but I am reasonably sure he did not.

Let us examine our two closest neighboring planets. The atmosphere of Mars is about one hundred times thinner than Earth's. One of the reasons is because Mars's atmosphere is 95 percent carbon dioxide. Here's a breakdown of Mars's composition, according to a NASA fact sheet[9]: Carbon dioxide is 95.32 percent; nitrogen is 2.7 percent; and other trace gases < 2 percent. Why is the CO_2 level in Mars's atmosphere so much higher than on Earth? Historically, humans never inhabited Mars. At present time, we have not sent people to Mars. Mars's CO_2 level is *more than* two thousand times higher than Earth's. So how did it get so polluted with CO_2? Afterall, there are no humans on Mars. Perhaps then, a planet can have high CO_2 levels without human contribution.

Now for the planet closest to Earth. The atmosphere of Venus is mostly carbon dioxide; its level a little higher than Mars's. The CO_2 level of the Venus atmosphere is 96.4 percent by volume, nearly 2,500 times that of Earth. Most of the remaining atmospheric gas is 3.4 percent nitrogen, with minimal amounts of other trace gases. Early evidence pointed to the sulfuric acid content in the atmosphere, but we now know that to be a rather minor constituent of the atmosphere.

So if human activity is the cause of increased CO_2 in the

---

[9] D. R. Williams, NASA Fact Sheet, Mars v Earth, NASA Goddard Space Flight Center, Sep. 27 2018, https://nssdc.gsfc.nasa.gov/planetary/factsheet/marsfact.html.

atmosphere, then what happened to all the humans on Mars and Venus? Where did they go?

Since the answer is obviously that humans never inhabited Mars or Venus, then why do so many believe greenhouse gases are the culprit for global warming? Because they have no idea as to the real reason, let alone understand it. They're sure the increase and decrease of solar energy cannot be the cause because they accept factors like these:

- Since climate data has been recorded, the average amount of energy coming from the sun is the same or has increased only slightly.
- If global warming was caused by the sun, there would be warmer temperatures in all layers of the atmosphere. Yet there seems to be a cooling in the upper atmosphere and a warming only at the lower atmosphere. They attribute that to greenhouse gases trapping in heat.
- Their climate models cannot reproduce the observed temperature trend over the past without artificially inserting a greenhouse gas effect.

All these HC[3] people preach carbon dioxide is the cause of climate change. I'm not sure why they promote this. Perhaps they all invest heavily in alternate energy sources.

When I was in grade school, I was taught CO_2 was a good thing. Humans and other animals exhale CO_2, and plants take in the CO_2 and expel oxygen. Do they still teach this, or has it changed? Humans and other animals need O_2, and the plants need CO_2, so it is a great symbiotic relationship. But climate change cultists say CO_2 is not only bad but it is downright poison. Why? Because CO_2 contains carbon. They believe carbon is so bad that they want us to only expel a minimal amount from our lungs. They even want to institute nationwide carbon credits. Their rationale is that since carbon is a poison, we can't allow any more into the environment, as our children will surly die.

Again, that is nonsense and far from the truth. CO_2 and carbon are actually good. Agriculture and the atmosphere benefit from it. Seems odd to me that those of the climate change leftist sect will pay more for organic produce than regular farmed produce. Yet *organic* means "relating to, or containing carbon."[10]

Okay, now on to methane (CH_4). The fanatics claim humans are responsible for the rise in methane gases. Atmospheric methane is a greenhouse gas but has remained relatively insignificant until now. Atmospheric methane is rising.

Is methane new? No, it's been around since the dawn of time. For the most part, its source is an offgas from the anaerobic decomposition of dead plants and animals. It's also expelled from dead vegetation in natural wetlands and aquatic farming such as rice paddy fields. To a lesser degree, CH_4 is also a resultant gas, including emissions from livestock production systems (including intrinsic fermentation of animal waste for manure); forest fires; anaerobic decomposition of organic waste in landfills; and some fossil methane emissions.

NASA scientists have calculated that over the past twenty years, the concentration of atmospheric methane has risen from around 1,770 ppb to·around 1,840 ppb. That's a whopping increase of seventy ppb—parts per billion! $\frac{70}{1,000,000,000}$ That's not very much.

Did you know methane is a good thing? It's used to power vehicles in many large cities. In fact, many landfill trucks use CH_4 to power them, and they obtain and use the methane captured from landfill release. In many countries, CH_4 is a primary

---

[10] *Merriam-Webster's Collegiate Dictionary*, s.v. "app (n.), organic.", 11th Edition 11th (eleventh), Merriam-Webster, Inc., Hardcover – 1994.

source of fuel to produce heat and light. Methane is also used to manufacture organic chemicals and other compounds such as the formulation of methanol (methyl alcohol) and is also a major fertilizer ingredient. In many Asian countries, methane is piped into homes for domestic heating and cooking. Germany annually produces enough electricity from burning methane to drive turbines to power on average 3.5 million homes. Many European factories, engines, and turbines are powered solely from the burning of methane. Further, methane is an excellent rocket fuel.

NASA says over the past twenty years the concentration of atmospheric methane has risen from around 1,770 ppb to around 1,840 ppb. I say good, we can use it.

Moreover, the leftists HC[3] sect have gone crazy on methane lately. New York City mayor Bill de Blasio declared all NYC public schools will have "meatless Mondays" beginning March 2019. His plan is to mandate that students eat only an all-vegetarian breakfast and lunch every Monday. Why? To quote Mayor de Blasio[11]: "Cutting back on meat a little will improve New Yorkers' health and reduce greenhouse gas emissions." Which greenhouse gas? Methane from cows. He calls his mandate "environmentally friendly meals."

Mayor de Blasio is supported on this notion by numerous other leftist nuts to include Congresswoman Alexandria Ocasio-Cortez (D-NY) and Senator Ed Markey (D-MA). Ocasio-Cortez has said[12], "Cows are 'public enemy number one' of our environmentalism movement." In her Green New Deal (GND), she says her ultimate aim is to transition Americans to a fully vegetarian diet in the

---

[11] S. Mcdonald, "NYC Mayor Bill De Blasio says its schools will have 'Meatless Mondays' this Fall, claiming it helps the Planet", Newsweek Magazine, Mar. 11, 2019.
[12] C. Hagle, "Six weeks of Fox's Alexandria Ocasio-Cortez obsession: "Totalitarian," "ignorant," "scary," and waging a "war on cows" ", Media Matters for America, Apr. 12, 2019.

name of halting apocalyptic climate change. That's right, halting it altogether. She has even suggested a terminal solution to methane-producing cattle.

This is misinformation. Did you know the methane from cows comes from their farts? How much can that really be? Other animals fart too.

In my research, I came across countless published papers and articles that were seemingly purposefully erroneous. For example, an article in *Science Daily*[13] stated, "Some greenhouse gases occur naturally in the atmosphere, while others result from human activities such as burning of fossil fuels such as coal." But the author never gave a single example of a human-caused greenhouse gas.

Another example: an article in the American Chemical Society publication *Chemistry for Life* similarly stated, "Some greenhouse gases occur naturally and enter the atmosphere as a result of both natural processes (such as decomposition of organic matter) and human activity (such as burning fossil fuels and agriculture)."[14]

While it is true humans burning stuff will release various gases into the atmosphere, humans did not really "create" these gases. Rather, they are formed as part of the natural cycle, no different from lightning striking a tree and starting a massive forest fire. Did a human cast down that lightning bolt? I think not.

To be even and fair, the article cited compounds that do not occur naturally and are, in fact, the result of human activity, including chlorofluorocarbons (CFCs), hydrochlorofluorocarbons (HCFCs), hydrofluorocarbons (HCFCs), bromofluorocarbons

---

[13] Umea University. (2018, September 4). Greenhouse emissions from Siberian rivers peak as permafrost thaws. *ScienceDaily*. Retrieved August 11, 2019 from www.sciencedaily.com/releases/2018/09/180904103229.htm.

[14] Chemistry for Life, American Chemical Society, ACS Climate Science Toolkit, Greenhouse Gases, Which Gases Are Greenhouse Gases?, https://www.acs.org/content/acs/en/climatescience/greenhousegases/whichgases.html

(halons), perfluorocarbons (PFCs), nitrogen trifluoride (NF_3), and sulfur hexafluoride (SF_6). But these compounds are so infinitesimal they can hardly be measured.

So if I am saying human beings releasing gases into the atmosphere is not the cause of global warming, then what is? The heat itself! In the next chapter, we will discuss what's causing temperatures to rise.

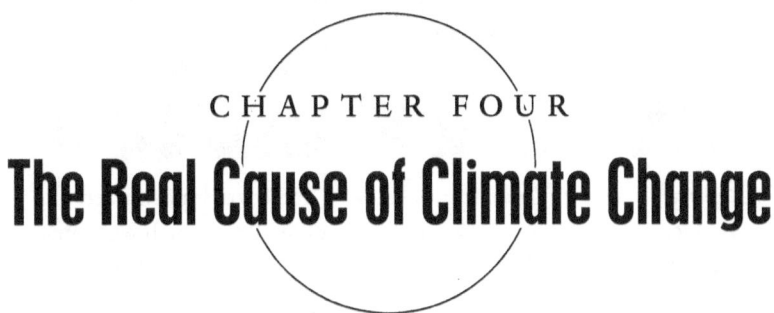

CHAPTER FOUR

The Real Cause of Climate Change

The earth is 92,955,807 miles from the sun at the center mass. At the equator, that distance changes little. But at the poles, the distance changes enough to give us the four seasons. That is because Earth is experiencing a wobble and is currently tilted 23.44 degrees. For simplicity, let's round it up to 23.5. This results in a variance of pole distance to the sun, which changes from summer solstice to winter solstice. This change in distance in the Northern Hemisphere from winter to summer results in an increase of about 6.9 percent in solar energy reaching the earth.

Now consider this (see figures 8 and 9): The diameter of Earth is 7,917.5 miles. The circumference of a sphere is equal to Π*D, or 3.1416 times the 7,917.5 miles. Therefore, the circumference of the earth is 24,900 miles. Since there are 360 degrees in a circle, each degree of Earth's surface equals 69.2 miles.

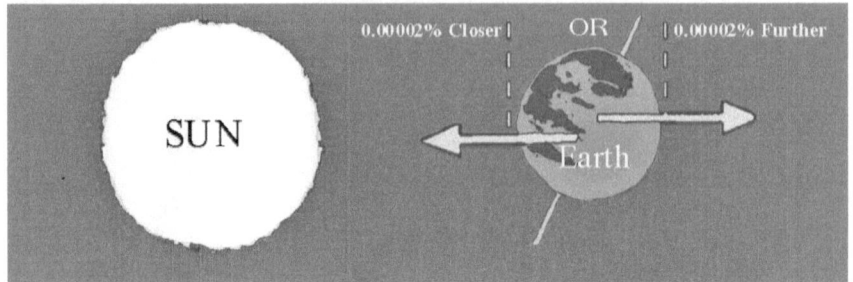

Figure 8: Source: Graphic by Jablecki | The Earth is positioned the perfect distance from the Sun to support life. If the Earth were any closer to, of further from the Sun, life as we know it would not exist.

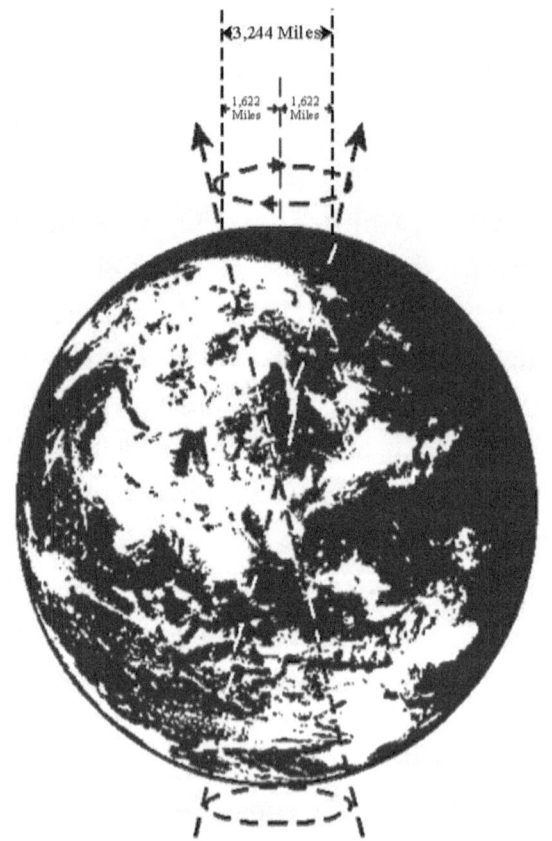

Figure 9: Source: Graphic by Jablecki | The Earth's tilt gives our planet 'normal' climate change. Summer in Northern Hemisphere while Winter in Southern Hemisphere.

So, at summer solstice the North Pole is 1,622 (69.2 X 23.44) miles closer to the sun, while at winter solstice, the North Pole is 1,622 (69.2 X 23.44) miles farther from the sun. The total variance is 3,244 miles. That 3,244 miles is the difference between winter cold and summer heat. If our planet were just 0.00002156 percent farther from the sun (two thousand miles), we would be much, much colder—*all the time*. Or, if our planet were just 0.00002156 percent closer to the sun, we would be much, much hotter—*all the time*. Do you think variance of 0.00002156 percent is very much? It's not. The earth is the perfect exact distance from the sun for life as we know it to exist.

When you were a kid, you probably played with a top or gyro. Remember how when you spun the top, it would stay stable no matter what you did to it (figure 10)?

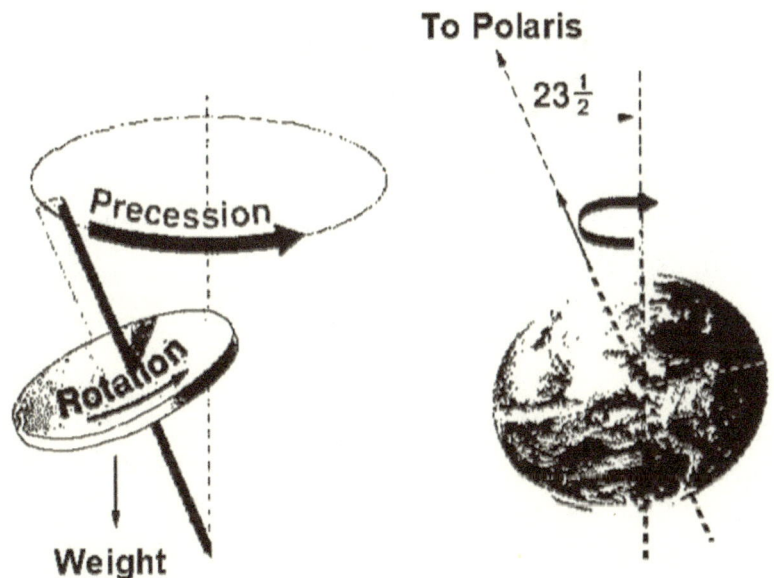

Figure 10: Source: Adapted from: Hean-Tatt, Ong., GUI World of Culture, Amazing Scientific World of Feng Shui, "Errors in Computation of Feng Shui and Astrological Years", January 2008. Axis of Earth's wobbles a full rotation every 26,000 years.

Even in your hand it remained stable. But as the speed of the top slowed, it started to wobble. With time, the wobble became greater and greater until the top began to drift around on the floor, eventually falling over.

The planets in our solar system are like that top. They spin fast, and at first, their high speed keeps them upright as they rotate around the sun. But like the top, as their speed degrades, they begin to wobble. Earth has been wobbling for a few million years. Right now, a complete single rotation of wobble takes about 26,000 years, which is a long time, so the wobble seems incredibly slow for us, but it's happening. We are constantly in a climatic change due to this wobble. As a result, the earth is at an incline. Whatever you call it, the earth is not straight up in relation to the orbit around the sun.

Many of you know the earth is at a tilt in relation to the sun. I suspect most of you knew nothing of the wobble, so you're about to have a solid explanation as to why Earth's climate is getting warmer.

For centuries, scientists have known that Earth's climate changes are caused by the wobbling rove of the planet while in orbit (figure 11).

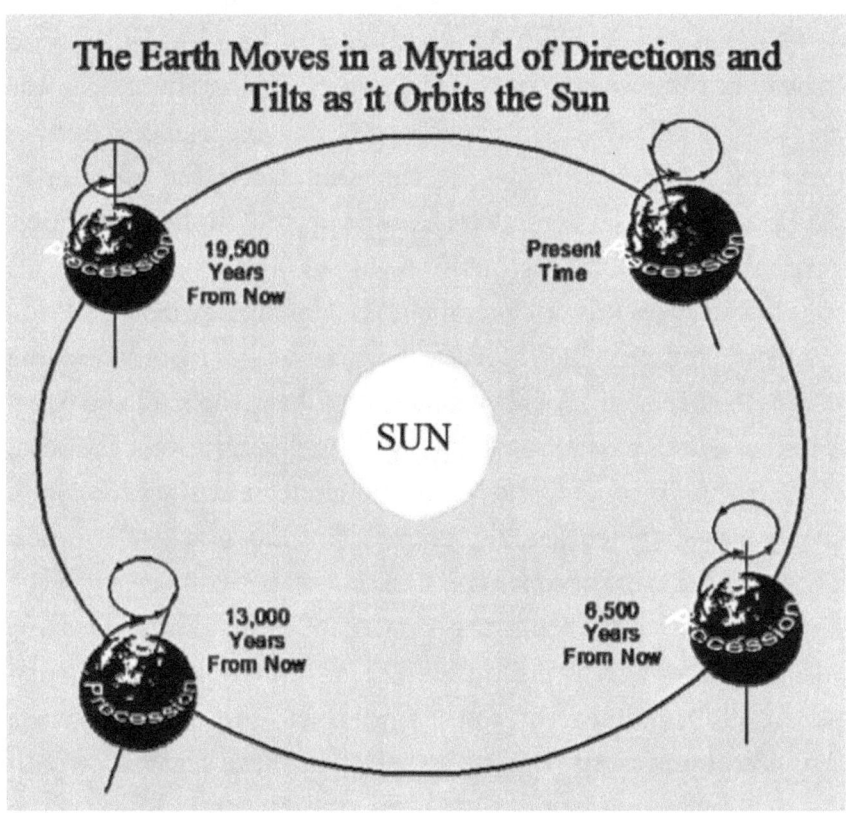

Figure 11: Source: Graphic by Jablecki via meta-composition of depiction in: Gregory Benson, Global Warming, Ice Ages, and Sea Level Changes: Something new or an astronomical phenomenon occurring in present day? What certain special interest groups don't want you to know., Jul. 10, 2010., https://sites.google.com/site/bensonfamilyhomepage/Home/ice-age-and-global-warming. With permission from Gregory Benson.

In 1899, the international latitude observatories were established for the purpose of measuring the wobble. They did their best, but their calculations were eventually superseded by other methods of measurement. Monitoring of the polar motion is now done by the International Earth Rotation Service (IERS).

What is the wobble? The wobble can best be understood by way of the example of the top or gyro losing energy, as was previously explained. This wobble the earth experiences changes the planets' orientation to the sun and affects the amount of sunlight and solar radiation reaching the higher latitudes, particularly the polar regions.

How do we know this? There is geologic evidence in soil samples taken deep below the surface, as well as from below the ocean floors. There is fossil evidence. Also, based on fruit and berry blossom periods from historical writings, including novels, the onset of the "spring" event has moved up about one week since the 1800s.

As was stated earlier, the diameter of Earth is 7,917.5 miles. This is important, because the size of a planet coupled with the distance of its orbit from the sun affect climate and weather. Also, we know Earth's circumference is 24,900 miles. During one solar cycle, as the earth moves around the sun, we experience four distinct seasons. But we experience these seasons only in the regions from north of the Tropic of Cancer and south of the Tropic of Capricorn. The region between the Tropic of Cancer and south of the Tropic of Capricorn to include the equator experiences very little seasonal change because this intertropical zone is not affected much by the earth's tilt.

At present time, the total variance from Earth's polar surface from summer to winter is 3,244 miles. That is not much variance considering how far Earth is from the sun. However, over time, as the wobble continues its twenty-six-thousand-year journey, not only does the relative angle change, the orbital distance expands and shrinks.[15] And that causes some of the climate shifts in both the Northern and Southern Hemispheres. So in the time human beings have been on Earth, we have inherited a period of somewhat neutral climate.

We know Earth is 92,955,807 miles from the sun, and only 3,244 miles will shift temperature from a winter daily average low temperature of five degrees Fahrenheit to a summer daily average high temperature of eighty-one degrees Fahrenheit, which is a seventy-six-degree Fahrenheit variance. That change is the result from distance variance of only 0.0034 percent. What if Earth were just one thousandth of one percent closer or farther from the sun? The planet would be uninhabitable for Homo sapiens and most likely all other animals. For just one thousandth of one percent (0.001 percent of 92,955,807 miles = 930 miles) is nearly a thousand miles.

But even greater forces are involved. Part of the wobble effect is the result solely of the axial rotation that takes thirteen thousand years to complete a cycle (figure 12). But an even greater shift is involved to explain the dramatic climate change. That wobble is the complete planet revolution about an unseen force establishing the central orbit.

Does this totally explain global climate change? It does not answer it all. In addition to the wobble, the earth experiences axial drift. During Earth's orbit, the gigantic forces generated cause the earth to shift in a circular fashion about its orbital path. The shift is slight but enough to cause significant changes in the global climate. At two times, the climate is what has been recently experienced by us living on the planet now, basically from 3000 BCE to 3000 CE, which is a six-thousand-year period or *orbital track quartile* (figure 13). On each side of this six-thousand-year cycle of "normal" climate is a transition period.

---

[15] "Expands and shrinks" is also often referred to as "waxes and wanes."

Earth's 26,000 Year Precessional Wobble

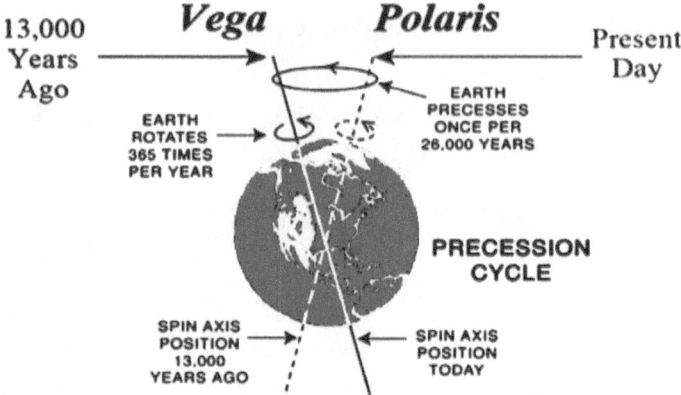

Figure 12: The Wobble gyrates in numerous directions. The wobble of just the axis to shift 180° takes 13,000 years. Source: Adapted by Jablecki from public domain and with Permission from Jordan Rabinowitz, Global Weather and Climate Center, LLC, in "The Milankovitch Cycle: The Difference Between Natural and Anthropogenic Heating (Credit: BBC)" 12/27/2018, https://www.globalweatherclimatecenter.com/climate-topics/the-milankovitch-cycle-the-difference-between-natural-and-anthropogenic-heating-credit-bbc.

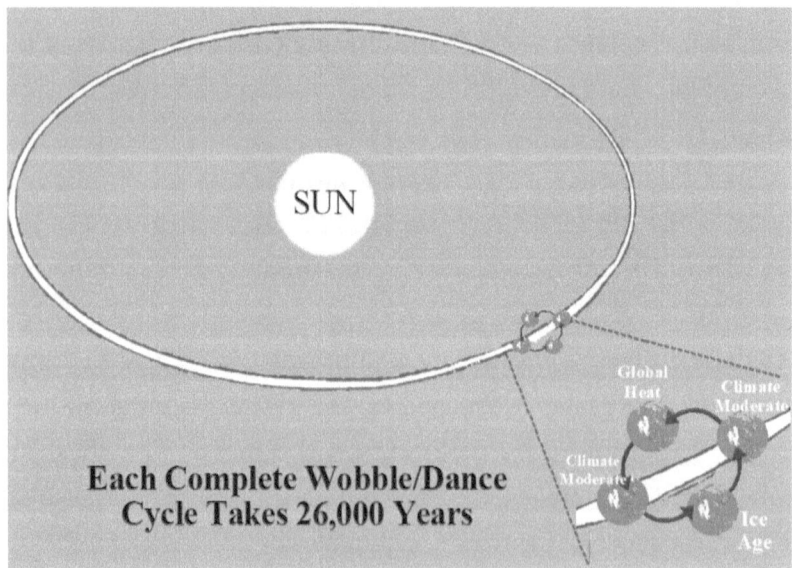

Figure 13: Source: Graphic by Jablecki | The Wobble gyrates in numerous directions. The greatest impact of climate is the Orbital Track Quartile.

Beginning about AD (or CE)[16] 3000, there will be a thousand-year period in which the climate transitions to hot. All global ice will melt and the earth will flood. Once the transition has completed, the earth's climate will be very, very hot. This period also will last about six thousand years and then there will be another thousand-year transition back to the moderate climate humans thrive in. This nice, "normal" climate will again last about six thousand years and then experience another thousand-year transition toward extreme cold. And the cycle continues.

Would Al Gore and those proponents supporting Mr. Gore's theory claim humans are responsible for some of the wobble in Earth's spin? I hope not. For that would be as ridiculous as thinking humans lowering CO_2 emissions will save Earth. No, there is nothing humans can do to stop this. It has been happening for a long time and will continue to happen for a long time to come.

Scientists have calculated, and have some proof, that since 1899, Earth's axis of spin has shifted about thirty-four feet. Now, research quantifies the reasons why and finds that a third is due to melting ice and rising sea levels, particularly in Greenland—placing the blame on the doorstep of anthropogenic climate change.

Another third of the wobble is due to land masses expanding upward as the glaciers retreat and lighten their load. The final portion is the fault of the slow churn of the mantle, the viscous middle layer of the planet.

Researchers have known that a proportion of this wobble was caused by glacial isostatic adjustment (GIA). GIA has been an ongoing process since the end of the last ice age. As the glaciers

---

[16] The age-old terms BC (Before Christ) and AD (Anno Domini, year of our Lord) are being replaced with the terms BCE and CE. This is being done as part of the anti-Christian movement's religiously neutral alternative. However, the new terms are anchored with the same "birth of Christ" yet avoid referring to Christ as Lord.

retreat, they relieve the land underneath their mass. Gradually, over thousands of years, the land responds to this relief by rising like bread dough. In some places on the edges of the ancient ice sheets, the land might also collapse because the ice had forced it to bulge upward.

But in new research recently published in *Earth and Planetary Science Letters*[17], Dr. Surendra Adhikari and his colleagues found that GIA was only responsible for about 1.3 inches (3.5 centimeters) of axis wobble per year. That roughly equates to only about a third of the wobble—four inches—observed each year over the twentieth century.

Furthermore, scientists have long known that the distribution of mass around the earth determines its spin, much like how the shape and weight distribution of a spinning top determines how it moves. Since the distribution of mass is not homogeneous, Earth's spin isn't perfectly even. Scientists know this because of observations in slight wiggles of the movements of the stars across the night sky that have been recorded for thousands of years. According to Erik Ivins, a study coauthor and a senior research scientist at Jet Population Laboratory (JPL), since the 1990s, space-based measurements have confirmed that the earth's axis of rotation drifts by a few centimeters a year, generally toward Hudson Bay in northeastern Canada.

Also reported in the 2018 study "What Drives Twentieth-Century Polar Motion?" published in *Earth and Planetary Science Letters*, the research team built a computer model of the physics of Earth's spin, feeding in data about changes in the balance of land-based ice and ocean waters over the twentieth century. The researchers accounted for other shifts in land and water, such as groundwater depletion and the building of artificial reservoirs, all part of humanity's terraforming of the planet.

---

[17] S. Adhikari et al., Earth and Planetary Science Letters, 502, pp126–132, 2018.

If you do your own research on Earth's shifting orbit, you'll find a great deal of theory but little proven fact. That includes the theory I propose in this book. You most likely will find the most discussion on the Milankovitch cycles. This theory was conceived by Serbian astronomer Milutin Milanković in the 1920s. He theorized that variations in eccentricity, axial tilt, and precession of Earth's orbit resulted in cyclical variation in the solar radiation reaching Earth. And it is this orbital forcing that strongly influences climatic patterns on Earth. My hypothesis is the same, except for the how and why. Milanković describes a complex interaction between orbital eccentricity,[18] axial tilt,[19] axial precession,[20] apsidal precession,[21] and orbital inclination,[22] a lot of complicated stuff that, perhaps because of its extreme complexity, has a lot of problems. My theory is much simpler and probable.

Science confirms that Earth's climate is getting warmer. The disagreement is as to how and why. Figure 14 depicts a graph of four major studies conducted by four different renowned institutions, which concur in findings.

Figure 15 depicts the Milankovitch cyclical movement related to Earth's orbit around the sun involving eccentricity, axial tilt, and precession.

---

[18] A parameter that determines the amount by which the Earth's orbit around the Sun deviates from a perfect circle. A value of 0 is a circular orbit, values between 0 and 1 form an elliptic orbit, 1 is a parabolic escape orbit, and greater than 1 is a hyperbola.

[19] The angle between the Earth's rotational axis and its' orbital axis, or, equivalently, the angle between its equatorial plane and orbital plane.

[20] Gravity-induced, slow, and continuous change in the orientation of the Earth's rotational axis.

[21] The precession (gradual rotation) of the line connecting the apsides (line of apsides) of the Earth's orbit.

[22] The tilt of the Earth in relation to the Sun expressed as the angle between a reference plane and the orbital plane or axis of direction of the Earth.

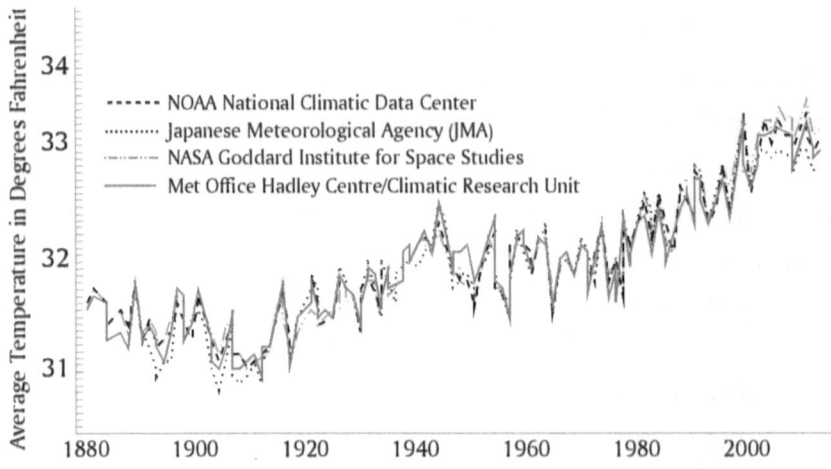

Figure 14: Source: Graphic by Jablecki | Global Average
Temperature 1880 - 2010 | Based on "Long-Term Global Warming
Trend Continues", The Earth Observatory is part of the EOS Project
Science Office at NASA Goddard Space Flight Center, 2013.

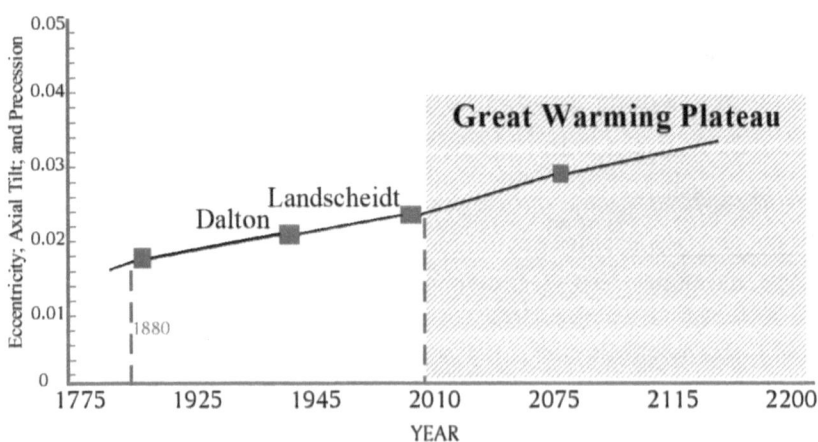

Figure 15: Source: Graphic by Jablecki | Milankovitch cyclical
movement related to the Earth's orbit around the Sun involving:
eccentricity; axial til; and precession. | Based on "2.5.2.4. Milankovitch
Cycles and Ice Ages", Global Climate Change, Blog, 2019.

Figure 16 depicts the merging of the collective scientific data on temperature warming (figure 14) and the Milankovitch cyclical movement (figure 15) for the same years (1800–2010).

The undeniable conclusion one must come to is that the "Heat Age" (global warming) will come. People can do nothing to prevent it. They can, however, prepare for it. We have a lot of time yet. Pinnacle warmth will be attained sometime after the year 3000. From now, 2019, until then, we can expect a mean (average) rise in temperature of about two degrees every one hundred years.

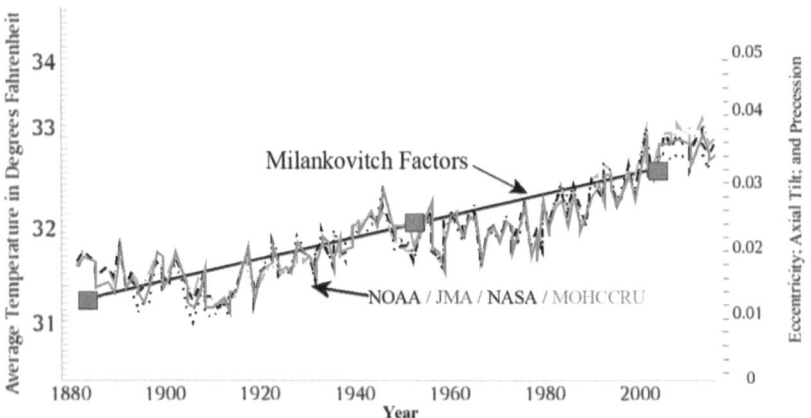

Figure 16: Source: Graphic by Jablecki | Milankovitch cyclical movement related to the Earth's orbit around the Sun {*Figure 14*} superimposed over the Global Average Temperature 1880 - 2010 {*Figure 15*}.

CHAPTER FIVE

Should We Worry?

I'm not going to worry about it, and I suggest you don't either. The people on planet Earth have nothing to worry about. For that matter, the next twenty to thirty generations don't have much to worry about either. But I'm sure some of you will worry anyway.

There is great controversy regarding polar ice melt. The science is not solid on this either. Many claim the ice at both poles is melting. Others claim the ice is not melting anywhere. Still others claim ice is melting at one pole but not the other.

Have you seen article titles like these: "National Oceanic and Atmospheric Administration (NOAA) Reports the Arctic Has Lost 95 Percent of Its Oldest, Thickest Ice" or "National Aeronautics and Space Administration Researchers Have Observed Ice Retreat in East Antarctica—a Region They'd Previously Believed Was Stable"?

Surely, you've seen articles like these. Is the information true? Who knows. But I think not. Most of the evidence indicates the ice at one pole is, in fact, melting at a faster rate than at the other pole.

The controversy with the South Pole (Antarctica) melt

surrounds scientific claims that Antarctic sea ice found near the South Pole has actually been increasing in recent years. Published research[23] has suggested Antarctic sea ice experienced its greatest growth between 2012 and 2014. How can that be if the climate is getting hotter? Well, it is not an even distribution because of all the complex twists and turns Earth goes through as it orbits the sun. Climate change deniers often point to this fact as proof that global warming is just a myth. But not so. Remember, I am not a climate change denier. I am sure it's happening (figure 17).

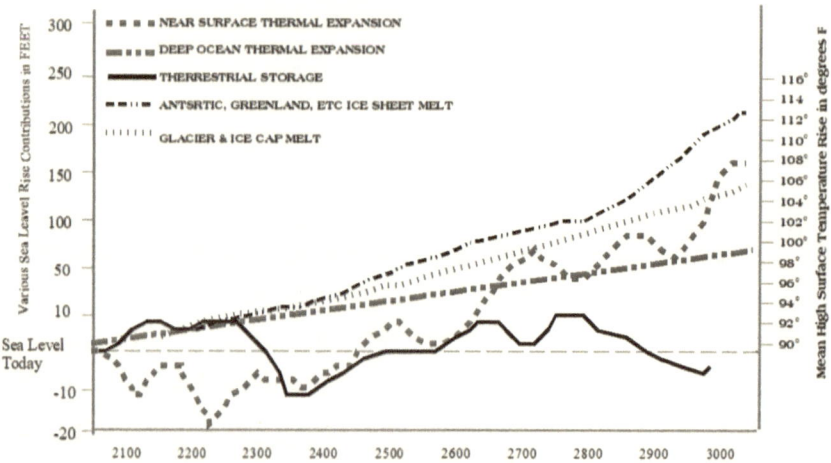

Figure 17: Source: Graphic by Jablecki | Based on observed sea level using coastal and island tide gauges and using TOPEX/Poseidon/Jason-1&2 satellite altimeter data, 1960-1020, Church, J. A. & White, N. J., "Sea-level rise from the late 19th to the early 21st century", *Surveys in Geophysics*, 2011.

[23] State of the Cryosphere, is the cryosphere sending signals about climate change?, National Snow and Ice Data Center, SOTC: Sea Ice, Nov. 20, 2018; S. E. Stammerjohn, et.al., Regions of rapid sea ice change: An inter-hemispheric seasonal comparison, Geophysical Research Ltrs. 39: L06501, 2012.; S. E. Stammerjohn, et. al., Seasonal sea ice changes in the Amundsen Sea, Antarctica, over the period of 1979-2014, Elementa: Sci. of the Anthropocene 3: 000055, 2015.; G. A. Meehl, et. al., Antarctic sea-ice expansion between 2000 and 2014 driven by tropical Pacific decadal climate variability. Nature Geoscience, 10.1038/NGEO2751, 2016.; Karl Mathiesen, Why is Antarctic sea ice at record levels despite global warming? The Guardian, Oct 9, 2014.

I simply disagree with the claim that human activities are the cause and choose to believe the facts of the past. It has happened and will continue to happen despite all of humankind's efforts.

What might explain this? Both poles are different. The North Pole (Arctic) is all ocean, no land. Additionally, it is somewhat enclosed and surrounded by land. Therefore, there is very little space for chunks of ice to float around, so the ice floes constantly collide into each other and pile up into thick ridges.

In contrast, the South Pole is a near geographic opposite of the North Pole. The Antarctic consists of land surrounded by an ocean. The sea ice is free to move and floats northward into warmer waters, where it eventually melts during the Southern Hemisphere's summer months.

When all the ice melts, and it will, what happens to the level of the oceans? As for the floating ice, it won't have much effect as most of the sea ice is already contributing to the level of the sea. You saw the movie *Titanic*, right? Most of the iceberg is underwater. So what about the ice on Antarctica, Greenland, Alaska, Russia ...?

Based on all the predictions I have reviewed, the rise in sea level is 210–240 feet! No clear single number is available. That will most certainly be a big rise, which will affect life on Earth. But this has happened before. How do you think all those fossilized shark teeth and fish bones got into the dirt in Colorado, Utah, Montana, Wyoming, and elsewhere?

Here's another phenomenon resulting from the wobble. Earth's magnetic pole is moving. Scientists say the earth's north magnetic pole has been shifting toward Siberia. The magnetic pole results from a massive concentration of molten iron within our planet's core. This molten mass is now sloshing around.

Does this mean that the North Pole may flip? This answer is suspected but not known. My guess is such a reversal will happen.

Yet there is currently no evidence of this imminent occurrence. That being said, the scientists at NASA say Earth's magnetic field has flipped its polarity many times over the millennia. Further, they say if you were alive about eight hundred thousand years ago and facing what we call north while holding a magnetic compass, the needle would actually point to the south.

Did you know the sun's magnetic field changes polarity approximately every eleven years? That's pretty often, and nothing bad has happened. The reversal occurs during each solar cycle as the sun's inner magnetic dynamo reorganizes itself. The sun's polar magnetic fields weaken, go all the way down to zero, and then increase again with opposite polarity. This is a regular part of the solar cycle.

The doomsday fanatics claim this next occurrence of Earth's polar reversal will lead to Earth's destruction. But would there really be any dramatic effects? The answer: based on all the geologic and fossil records we have from literally hundreds of previous magnetic polarity reversals, and like the sun, there will be no significant effect. So, again, nothing to worry about.

Earth's atmosphere has been what it is for a billion years. It will most likely remain such for a billion more, the exception being a cataclysmic event such as the collision of a large meteor, asteroid, or comet. Even with such an event, the atmosphere would repair itself with time.

One might say, "If only you'd look at the evidence, you'd understand." But unfortunately, a great many people just look past the evidence to hang on to their preconceived beliefs. To paraphrase Dr. Thomas Sowell, It is only when you go beyond rhetoric, and start looking at hard facts, that your belief becomes a big disappointment, if not a disaster.

In summary, we have at least a thousand years before the temperature on Earth gets too hot to bear. A great deal can be done in preparation between now and then. We should not worry about what is to come; rather, we should prepare for inevitable extreme heat. Preparation! Yes, much time should be spent on preparation, and one cannot imagine what technology will bring. As an example, we can build large cities with domes over them to capture the solar radiation and convert that energy to a cool, comfortable, and controlled climate. Another example: we can build underground cities or communities within mountains. The insulation qualities of the rock and soil would be so beneficial that we could easily cool these communities to an adequate temperature. The downside is that we would have to live that way for about five thousand years because it will take about that long for the wobble to rotate again into the comfortable zone in which human life thrives.

Remember, humans have accomplished a great deal in the past thousand years. No one can possibly imagine what scientific industrial knowledge will come to fruition between now and a thousand years from now.

APPENDIX A

Data Reported to Carbon Dioxide Information Analysis Center, Environmental Sciences Division, Oak Ridge National Laboratory, Tennessee, USA.

CO_2 Emissions (Metric Tons per Capita) Source: The World Bank

CO2 emissions (metric tons per capita)				
	1960	Ranking	2014	Ranking
Afghanistan	0.05	157	0.3	202
Albania	1.26	58	1.98	134
Algeria	0.55	88	3.72	105
American Samoa		T- 173		T- 229
Andorra		T- 173	5.83	72
Angola	0.1	143	1.29	156
Antigua and Barbuda	0.66	81	5.38	77
Arab World	0.64	84	4.86	84
Argentina	2.37	40	4.75	86
Armenia		T- 173	1.9	139
Aruba		T- 173	8.41	39
Australia	8.58	9	15.37	16
Austria	4.37	24	6.87	53
Azerbaijan		T- 173	3.93	103
Bahamas, The	3.75	27	6.32	59
Bahrain	3.54	32	23.45	5
Bangladesh		T- 173	0.46	191
Barbados	0.75	78	4.49	91

CO2 emissions (metric tons per capita)				
	1960	Ranking	2014	Ranking
Belarus		T- 173	6.7	54
Belgium	9.94	8	8.33	40
Belize	0.48	92	1.41	153
Benin	0.07	152	0.61	184
Bermuda	3.55	31	8.84	37
Bhutan		T- 173	1.29	157
Bolivia	0.27	111	1.93	137
Bosnia and Herzegovina		T- 173	6.23	61
Botswana		T- 173	3.24	114
Brazil	0.65	83	2.59	124
British Virgin Islands		T- 173	6.07	67
Brunei Darussalam	4.08	25	22.12	7
Bulgaria	2.83	38	5.87	71
Burkina Faso	0.01	170	0.16	215
Burundi		T- 173	0.04	228
Cabo Verde	0.11	141	0.93	168
Cambodia	0.04	160	0.44	192
Cameroon	0.05	158	0.31	200
Canada	10.77	7	15.12	17
Caribbean small states	1.4	54	8.88	36
Cayman Islands	1.4	55	9.17	31
Central African Republic	0.06	154	0.07	222
Central Europe and the Baltics	5.1	21	6.15	65
Chad	0.02	167	0.05	227
Channel Islands		T- 173		T- 229
Chile	1.75	44	4.69	88
China	1.17	59	7.54	45
Colombia	1	66	1.76	143
Comoros	0.06	155	0.2	212
Congo, Dem. Rep.	0.15	131	0.06	225
Congo, Rep.	0.22	122	0.64	181
Costa Rica	0.37	99	1.63	147
Cote d'Ivoire	0.13	136	0.49	190
Croatia		T- 173	3.97	101
Cuba	1.92	43	3.05	116
Curacao		T- 173	37.73	2

CO2 emissions (metric tons per capita)				
	1960	Ranking	2014	Ranking
Cyprus	1.55	50	5.26	80
Czech Republic		T- 173	9.17	32
Denmark	6.5	16	5.94	70
Djibouti	0.48	93	0.79	178
Dominica	0.18	126	1.86	140
Dominican Republic	0.32	100	2.07	132
East Asia & Pacific	1.16	61	6.29	60
East Asia & Pacific (excluding high income)	0.95	69	5.78	73
Ecuador	0.39	97	2.76	121
Egypt, Arab Rep.	0.59	87	2.2	130
El Salvador	0.22	123	1	167
Equatorial Guinea	0.09	145	4.73	87
Eritrea		T- 173		T- 229
Estonia		T- 173	14.85	18
Eswatini	0.09	146	0.93	169
Ethiopia	0.02	168	0.12	219
Euro area		T- 173	6.47	56
Europe & Central Asia		T- 173	6.92	52
Europe & Central Asia (excluding high income)		T- 173	7.42	48
European Union	5.76	19	6.38	58
Faroe Islands	1.69	47	12.24	22
Fiji	0.49	91	1.32	155
Finland	3.41	33	8.66	38
Fragile and conflict affected situations	0.28	110	0.89	171
France	5.79	18	4.57	90
French Polynesia	0.47	95	2.92	118
Gabon	0.26	114	2.77	120
Gambia, The	0.05	159	0.27	208
Georgia		T- 173	2.41	126
Germany		T- 173	8.89	35
Ghana	0.22	124	0.54	189
Gibraltar	2.04	42	15.51	14
Greece	1.13	63	6.18	63
Greenland	6.88	13	8.99	33

CO2 emissions (metric tons per capita)				
	1960	Ranking	2014	Ranking
Grenada	0.24	117	2.28	128
Guam		T- 173		T- 229
Guatemala	0.32	101	1.15	159
Guinea	0.11	142	0.21	211
Guinea-Bissau	0.03	165	0.16	214
Guyana	1.15	62	2.63	122
Haiti	0.07	153	0.27	207
Heavily indebted poor countries (HIPC)	0.12	139	0.27	206
Honduras	0.3	106	1.08	162
Hong Kong SAR, China	0.96	68	6.39	57
Hungary	4.54	23	4.27	99
Iceland	6.91	12	6.06	68
India	0.27	112	1.73	145
Indonesia	0.24	118	1.82	141
Iran, Islamic Rep.	1.71	45	8.28	41
Iraq	1.13	64	4.81	85
Ireland	3.95	26	7.31	50
Isle of Man		T- 173		T- 229
Israel	3.06	35	7.86	43
Italy	2.18	41	5.27	79
Jamaica	0.9	72	2.59	123
Japan	2.52	39	9.54	27
Jordan	0.8	77	3	117
Kazakhstan		T- 173	14.36	19
Kenya	0.3	107	0.31	197
Kiribati		T- 173	0.56	187
Korea, Dem. People's Rep.		T- 173	1.61	148
Korea, Rep.	0.5	90	11.57	24
Kosovo		T- 173		T- 229
Kuwait	28.94	2	25.22	4
Kyrgyz Republic		T- 173	1.65	146
Lao PDR	0.04	161	0.3	203
Latin America & Caribbean	1.32	57	3.06	115
Latin America & Caribbean (excluding high income)	1.17	60	2.78	119

CO2 emissions (metric tons per capita)				
	1960	Ranking	2014	Ranking
Latvia		T- 173	3.5	110
Least developed countries: UN classification	0.1	144	0.31	199
Lebanon	1.43	53	4.3	97
Lesotho		T- 173	1.15	160
Liberia	0.15	132	0.21	210
Libya	0.48	94	9.19	30
Liechtenstein		T- 173	1.19	158
Lithuania		T- 173	4.38	94
Luxembourg	36.69	1	17.36	10
Macao SAR, China	0.31	104	2.18	131
Macedonia, FYR		T- 173	3.61	107
Madagascar	0.08	149	0.13	216
Malawi		T- 173	0.07	224
Malaysia		T- 173	8.03	42
Maldives		T- 173	3.27	113
Mali	0.02	169	0.08	221
Malta	1.04	65	5.4	76
Marshall Islands		T- 173	1.94	136
Mauritania	0.04	162	0.67	180
Mauritius	0.27	113	3.35	112
Mexico	1.65	48	3.87	104
Micronesia, Fed. Sts.		T- 173	1.45	152
Middle East & North Africa	0.97	67	6.16	64
Middle East & North Africa (excluding high income)	0.86	74	3.93	102
Moldova		T- 173	1.39	154
Monaco		T- 173		T- 229
Mongolia	1.35	56	7.13	51
Montenegro		T- 173	3.36	108
Morocco	0.3	108	1.74	144
Mozambique	0.26	115	0.31	198
Myanmar	0.13	137	0.42	193
Namibia		T- 173	1.58	150
Nauru		T- 173	4.02	100
Nepal	0.01	171	0.28	205

CO2 emissions (metric tons per capita)				
	1960	Ranking	2014	Ranking
Netherlands	6.4	17	9.92	26
New Caledonia	10.91	6	16.01	13
New Zealand	4.87	22	7.69	44
Nicaragua	0.3	109	0.81	176
Niger	0.01	172	0.11	220
Nigeria	0.08	150	0.55	188
North America	15.53	4	16.35	12
Northern Mariana Islands		T- 173		T- 229
Norway	3.66	29	9.27	29
OECD members	7.26	10	9.52	28
Oman		T- 173	15.44	15
Other small states	0.63	85	7.52	47
Pacific island small states	0.38	98	1.06	163
Pakistan	0.32	102	0.9	170
Palau	1.52	51	12.34	21
Panama	0.88	73	2.25	129
Papua New Guinea	0.09	147	0.81	177
Paraguay	0.16	129	0.87	173
Peru	0.81	76	1.99	133
Philippines	0.32	103	1.06	164
Poland	6.74	14	7.52	46
Portugal	0.93	70	4.33	95
Puerto Rico		T- 173		T- 229
Qatar	3.71	28	45.42	1
Romania	2.9	37	3.52	109
Russian Federation		T- 173	11.86	23
Rwanda	0.04	163	0.07	223
Samoa	0.14	133	1.03	165
San Marino		T- 173		T- 229
Sao Tome and Principe	0.17	128	0.59	185
Saudi Arabia	0.66	82	19.53	8
Senegal	0.26	116	0.61	183
Serbia		T- 173	5.28	78
Seychelles		T- 173	5.42	75
Sierra Leone	0.31	105	0.18	213
Singapore	0.85	75	10.31	25

CO2 emissions (metric tons per capita)				
	1960	Ranking	2014	Ranking
Saint Maarten (Dutch part)		T- 173	19.46	9
Slovak Republic		T- 173	5.66	74
Slovenia		T- 173	6.21	62
Small states	0.92	71	7.38	49
Solomon Islands	0.09	148	0.35	196
Somalia	0.03	166	0.05	226
South Africa	5.61	20	8.98	34
South Asia	0.24	119	1.46	151
South Sudan		T- 173	0.13	218
Spain	1.61	49	5.03	81
Sri Lanka	0.23	120	0.89	172
St. Kitts and Nevis	0.21	125	4.3	98
St. Lucia	0.16	130	2.31	127
St. Martin (French part)		T- 173		T- 229
St. Vincent and the Grenadines	0.14	134	1.91	138
Sub-Saharan Africa	0.55	89	0.84	175
Sudan	0.13	138	0.3	201
Suriname	1.49	52	3.63	106
Sweden	6.58	15	4.48	93
Switzerland	3.66	30	4.31	96
Syrian Arab Republic	0.7	79	1.6	149
Tajikistan		T- 173	0.62	182
Tanzania	0.08	151	0.22	209
Thailand	0.14	135	4.62	89
Timor-Leste		T- 173	0.39	194
Togo	0.04	164	0.36	195
Tonga	0.18	127	1.14	161
Trinidad and Tobago	3.04	36	34.16	3
Tunisia	0.41	96	2.59	125
Turkey	0.61	86	4.49	92
Turkmenistan		T- 173	12.52	20
Turks and Caicos Islands		T- 173	6.09	66
Tuvalu		T- 173	1.01	166
Uganda	0.06	156	0.13	217
Ukraine		T- 173	5.02	82

CO2 emissions (metric tons per capita)				
	1960	Ranking	2014	Ranking
United Arab Emirates	0.12	140	23.3	6
United Kingdom	11.15	5	6.5	55
United States	16	3	16.49	11
Uruguay	1.7	46	1.97	135
Uzbekistan		T- 173	3.42	111
Vanuatu		T- 173	0.59	186
Venezuela, RB	7.01	11	6.03	69
Vietnam	0.23	121	1.8	142
Virgin Islands (U.S.)		T- 173		T- 229
West Bank and Gaza		T- 173		T- 229
World	3.1	34	4.97	83
Yemen, Rep.	0.7	80	0.86	174
Zambia		T- 173	0.29	204
Zimbabwe		T- 173	0.78	179

SUGGESTED READING

Adhikari, S., L. Caron, B. Steinberger, J. T. Reager, K. K. Kjeldsen, B. Marzeion, E. Larour, and E. R. Ivins. "What Drives Twentieth Century Polar Motion?" *Earth and Planetary Science Letters* 502: 126–32. doi: 10.1016/j.epsl.2018.08.059.

Allen, C. D., A. K. Macalady, H. Chenchouni, et al. "A Global Overview of Drought and Heat-Induced Tree Mortality Reveals Emerging Climate Change Risks for Forests." *Forest Ecology and Management* (2010).

Allison, I. et al. "The Copenhagen Diagnosis: Updating the World on the Latest Climate Science." (Sydney: UNSW Climate Change Research Center, 2009).

Biello, D. "CO_2 Levels for February Eclipsed Prehistoric Highs." *Scientific American*, 2015.

Church, J. A. and N. J. White. "A Twentieth-Century Acceleration in Global Sea Level Rise." *Geophysical Research Letters* 33, L01602 (2006).

Cook, J., D. Nuccitelli, S. A. Green, et al. "Quantifying the Consensus on Anthropogenic Global Warming in the Scientific Literature." *Environmental Research Letters* 8 (2013).

Derksen, C. and R. Brown. "Spring Snow Cover Extent Reductions in the 2008–2012 Period Exceeding Climate Model Projections." *Geophysical Research Letters* 39, L19504 (2012).

Ding D., E. W. Maibach, X. Zhao, et al. "Support for Climate Policy and Societal Action Are Linked to Perceptions about Scientific Agreement." *Nature Climate Change* 2011; DOI: 10.1038/nclimate1295.

Hamilton, L. C. "Public Awareness of the Scientific Consensus on Climate." (SAGE Open, 2006).

———"Statistics with Stat, Version 12." (Belmont CA: Cengage, 2013).

Hamilton, L. C., J. Hartter, B. D. Keim, et al. "Wildfire, Climate, and Perceptions in Northeast Oregon." *Regional Environmental Change* (2016).

Hamilton, L. C., J. Hartter, F. R. Stevens, et al. "Forest Views: Shifting Attitudes toward the Environment in Northeast Oregon." (Carsey Institute, 2012).

Hamilton, L. C., J. Hartter, M. Lemcke-Stampone, et al. "Tracking Public Beliefs about Anthropogenic Climate Change." *PLOS ONE* 10, no. 9 (2015).

Hamilton, L. C., J. Hartter, T. G. Safford, and F. R. Stevens. "Rural Environmental Concern: Effects of Position, Partisanship and Place." *Rural Sociology* (2014).

Kahan, D. M. "Climate Science Communication and the Measurement Problem." *Advances in Political Psychology* (2015).

Kwok, R. and D. A. Rothrock. "Decline in Arctic Sea Ice Thickness from Submarine and ICESAT Records: 1958–2008." *Geophysical Research Letters* 36 (2009).

Lean, J. "Cycles and Trends in Solar Irradiance and Climate." *Wiley Interdisciplinary Reviews: Climate Change* 1 (2010).

Levitus, S., J. I. Antonov, T. P. Boyer, R. A. Locarnini, H. E. Garcia, and A. V. Mishonov. "Global Ocean Heat Content 1955–2008 in Light of Recently Revealed Instrumentation Problems." *Geophysical Research Letters* 36, L07608 (2009).

Lockwood, M. "Solar Change and Climate: An Update in the Light of the Current Exceptional Solar Minimum." *Proceedings of the Royal Society A* (December 2, 2009).

Maibach, E., T. Myers, and A. Leiserowitz. "Climate Scientists Need to Set the Record Straight: There Is a Scientific Consensus That Human-Caused Climate Change Is Happening." *Earth's Future* (2014).

Melillo, J. M., T. C. Richmond, and G. W. Yohe, eds. "Climate Change Impacts in the United States: The Third National Climate Assessment." (Washington, DC: US Global Change Research Program, 2014).

Mueller, I. I. *Spherical and Practical Astronomy as Applied to Geodesy* (New York: Frederick Ungar Publishing, 1969).

National Research Council (NRC). "Surface Temperature Reconstructions for the Last Two Thousand Years." (Washington, DC: National Academy Press, 2006).

Oreskes, N. "The Scientific Consensus on Climate Change." *Science* 306 (2004).

Peterson, T. C. et.al. "State of the Climate in 2008." Supplement, *Bulletin of the American Meteorological Society* 90, no. 8 (August 2009).

Polyak, L. et al. "History of Sea Ice in the Arctic." (US Geological Survey, Climate Change Science Program Synthesis and Assessment Product, 2009).

Ramaswamy, V. et al. "Anthropogenic and Natural Influences in the Evolution of Lower Stratospheric Cooling." *Science* 311 (2006).

Santer, B. D. et al. "A Search for Human Influences on the Thermal Structure of the Atmosphere." *Nature* 382 (1996).

Van der Linden, S. L., A. A. Leiserowitz, G. D. Feinberg, and E. W. Maibach. "How to Communicate the Scientific Consensus on Climate Change: Plain Facts, Pie Charts, or Metaphors?" *Climate Change* 126 (2014).

www.ingramcontent.com/pod-product-compliance
Lightning Source LLC
Chambersburg PA
CBHW021502210526
45463CB00002B/852